Lecture Notes on Mathematical Modelling in the Life Sciences

The rapid pace and development of new methods and techniques in mathematics and in biology and medicine creates a natural demand for up-to-date, readable, possibly short lecture notes covering the breadth and depth of mathematical modelling, mathematical analysis and numerical computations in the life sciences, at a high scientific level.

The volumes in this series are written in a style accessible to graduate students. Besides monographs, we envision the series to also provide an outlet for material less formally presented and more anticipatory of future needs due to novel and exciting biomedical applications and mathematical methodologies.

The topics in LMML range from the molecular level through the organismal to the population level, e.g. gene sequencing, protein dynamics, cell biology, developmental biology, genetic and neural networks, organogenesis, tissue mechanics, bioengineering and hemodynamics, infectious diseases, mathematical epidemiology and population dynamics.

Mathematical methods include dynamical systems, partial differential equations, optimal control, statistical mechanics and stochastics, numerical analysis, scientific computing and machine learning, combinatorics, algebra, topology and geometry, etc., which are indispensable for a deeper understanding of biological and medical problems.

Wherever feasible, numerical codes must be made accessible.

Founding Editors:

Michael C. Mackey, McGill University, Montreal, QC, Canada

Angela Stevens, University of Münster, Münster, Germany

Yoichiro Mori • Benoît Perthame • Angela Stevens
Editors

Dynamics of Physiological Control

Contributions in Honor of Michael C. Mackey

 Springer

Editors
Yoichiro Mori
Departments of Mathematics and Biology
University of Pennsylvania
Philadelphia, PA, USA

Benoît Perthame
Laboratoire Jacques-Louis Lions
Sorbonne Université
Paris, France

Angela Stevens
Applied Mathematics, Institute for Analysis
and Numerics
University of Münster
Münster, Germany

ISSN 2193-4789   ISSN 2193-4797   (electronic)
Lecture Notes on Mathematical Modelling in the Life Sciences
ISBN 978-3-031-82395-4   ISBN 978-3-031-82396-1   (eBook)
https://doi.org/10.1007/978-3-031-82396-1

This Springer imprint is published by the registered company Springer Nature Switzerland AG
The registered company address is: Gewerbestrasse 11, 6330 Cham, Switzerland

If disposing of this product, please recycle the paper.

Preface

This volume of the Lecture Notes on Mathematical Modeling in the Life Sciences (LMML) is dedicated to Michael Mackey on the occasion of his 80th birthday and is finally published on the occasion of his 82nd birthday.

It is a great pleasure for us to present this collective enterprise.

Michael Mackey is one of the two founders of LMML. With a Bachelor in Mathematics and a PhD in Physiology and Biophysics, he explored the promising and strong connections between mathematical modeling and medicine and biology early on.

His fields of research and his motto:

> "My emphasis is always on biologically realistic mathematical models, careful consideration of laboratory and/or clinical data, and achieving a reasonable concordance between the data and the modeling."

always were and still are central for this series.

Michael Mackey is consistently eager to discuss science concretely and critically, which is nicely reflected also in the contributions to this volume.

These Lecture Notes start with some personal notes about Michael Mackey's career and his research by David Dale, followed by Morgan Craig's article about data-driven models of chemotherapy-induced neutropenia, further explaining Michael Mackey's contributions and his collaborators' toward understanding this side effect of cytotoxic chemotherapy.

Then Albert Goldbeter stresses the ubiquitousness of oscillations in biological systems, and how multiple layers of regulation by feedback loops and the cooperativity of biological processes provide the sources of nonlinearity responsible for the onset of oscillatory behavior.

In his contribution about mathematical models of heterogeneous stem cell regeneration, Jinzhi Lei focuses on primary strategies for describing cell division in biological systems and on methods for modeling Waddington's epigenetic landscape. These are fundamental steps to understand tissue development and tumor progression.

Pauline Mazel, Nicolas Foray, and Laurent Pujo-Menjouet describe the effect of irradiation and antioxidants in cells affected by Alzheimer's disease, focusing on ATM kinase as a pivotal protein in the cellular response to genotoxic stress, playing a crucial role in the recognition and repair of DNA double-strand breaks.

In his article about ergodic and chaotic properties in biological models Ryszard Rudnicki discusses two classes of mathematical models, one describing, e.g., the change of population size in successive generations, and the other being partial differential equations, e.g. space-structured models or describing maturity-distribution of precursors of blood cells.

Moisés Santillán reviews mathematical models for quantitative insight into glucose regulation and addresses new challenges such as integrating glucose regulation into whole-body regulatory networks and unraveling long-term compensatory mechanisms.

Last but not least, John Milton and Tamás Insperger explain why a stick balanced at the fingertip falls, although from the mathematical point of view, the upright position of an inverted pendulum can be fully stabilized by time-delayed feedback.

This volume would not have been possible without the essential work of referees, to whom we are very grateful.

Here's to you Mike, in recognition of your extra-ordinary impact on the "mathematics in the life sciences" community.

We are all wishing you many more healthy and happy years to come!

Philadelphia, PA, USA   Yoichiro Mori
Paris, France   Benoît Perthame
Münster, Germany   Angela Stevens
November 2024

Contents

Mike and Me Since 1973: A Review of a Long Friendship and Research Relationship

David C. Dale

Michael Mackey and I have known each other for a long time! He was a graduate student and received his Ph.D. in Physiology from the University of Washington (UW) in 1968, now my academic home. Mike reports that he got his inspiration to study the dynamics of cell growth at the UW from a fellow graduate student, Dr. John Combs [1]. We both began working at the National Institutes of Health (NIH) in 1968 but we did not know each other then. At that time the US had an ongoing universal military draft. I, like Mike, saw benefit for my family and my career if I if I got a job at the NIH rather than going to the war in Vietnam. At the NIH Mike began studying the classic cell cycle, focusing on the rodent mammary tumor model as proposed by Burns and Tannock [2, 3].

I landed a job at the NIH Clinical Center in the Laboratory of Clinical Investigation of the National Institute of Allergy and Infectious Diseases (NIAID) under its new director, Dr. Sheldon Wolff. Shelly, as we all called him, was a visionary person who wanted to investigate every reason people have for susceptibility to infections, i.e., diseases that are often associated with recurrent fevers. One of my first patients had a rare disease that causes recurrent fevers, a disease called "cyclic neutropenia." One of the early questions was how to be sure about the diagnosis. I thought, if we just had enough serial blood counts, we could know if there was a neutrophil cycle. Because we had research beds and could invite patients to stay for long periods of time, we asked patients to stay for several weeks at the NIH Clinical Center to observe their health and have blood samples drawn at the same time every day to see if there was a regular pattern of fluctuations in their blood cell counts. I was interested in mathematics and knew a little, but I recognized that I would need help in analyzing the data to look for periodicity. At that time, I was lucky to have made

D. C. Dale (✉)
Department of Medicine, University of Washington, Seattle, WA, USA
e-mail: dcdale@uw.edu

© The Author(s), under exclusive license to Springer Nature Switzerland AG 2025   1
Y. Mori et al. (eds.), *Dynamics of Physiological Control*, Lecture Notes
on Mathematical Modelling in the Life Sciences,
https://doi.org/10.1007/978-3-031-82396-1_1

a lasting friendship with the Institute's statistician, Dr. David Alling. He taught me about periodicity and the Lomb periodogram, and we began to work together.

It was quite fortuitous that at just about this time, researchers at Washington State University discovered that grey collie dogs, purebred collies with a grey coat rather than a true brown or black, died soon after birth due to infections [4]. Serial blood cell counts showed that there was an apparent cycle of blood neutrophils and the severe infections always coincided with severe neutropenia [4]. Shelly helped me find collie breeders who were concerned about the dying grey puppies. We created a home for these puppies at the NIH, and I became their personal physician! I cared for them with the help of many other physicians, researchers, veterinarians, and technicians and studied their disease as a model for human cyclic neutropenia for the next 40 years [5, 6]! One distinctive feature of the collies' cyclic neutropenia was the cycle length. In humans it was almost always about every 21 days, however, in the grey collies it was every 13 days [7, 8]. Bone marrow studies showed that there were regular periodic changes in the formation of neutrophils in the bone marrow in both the human and the canine form of the disease. Consistently, the marrow showed periodic interruption followed by recovery in neutrophil production [7, 8]. We started to ponder what was interrupting neutrophil production.

In Mike's paper "The story of the Mackey-Glass equation," he wrote that in 1974 John Combs pointed out to him our reports about human and canine cyclic neutropenia and another about periodic leukemia. Mike reported, "A wise suggestion as it turned out, since the study of these periodic hematological diseases became a focal point of my career" [1]. At that time, I was also trying to measure the bone marrow production of neutrophils in several ways: serial biopsies, labeling cells with radioisotopes, looking for the slope of the decline of counts to their lowest levels and then the pattern of recovery in this interesting disease. I read and reread the reports from the 1950s and the early 1960s on leukocyte kinetics by Cronkite et al. [9], Craddock et al. [10], and Athens et al. [11]. I became convinced that neutropenia was caused by an interruption in neutrophil production at an early stage in myelopoiesis. We observed that not only neutrophils, but red cells, monocytes and platelets also cycled, though not so extremely as the neutrophils. For a time, we referred to the disease as cyclic hematopoiesis, especially after we showed that the grey collies' disease could be cured by bone marrow transplantation [5, 7, 12].

Two important discoveries shifted our thinking and work. Mike and Leon Glass were then studying oscillatory dynamics. In modeling of cyclic neutropenia, Mike postulated that the fundamental problem was an inherent delay in the feedback mechanism regulating neutrophil production that caused cycling [13]. This paper in the journal *Blood* was a critical "paradigm shifting" report for understanding biological cycling. At the same time that Mike and Leon Glass were developing these concepts, I was interested in the newly discovered growth factors that could stimulate blood cell formation in the laboratory; for leukocytes these were called "colony-stimulating factors (CSFs)." We set out to measure the colony-stimulating factor levels in blood and urine of the grey collies and humans with cyclic neutropenia. We observed that the levels were highest when the marrow was recovering from severe neutropenia and did not rise until there was neutropenia [14, 15].

Normally the body is steadily overproducing neutrophils as an always ready supplier of microbicidal substances to fight infections anywhere in the body, "a handy infection-controlling fire department." However, when production is interrupted from the early stages of myelopoiesis there could be a lag in the signal to simulate production until the bone marrow supply of these cells declines to a critical level. Based on our studies in the grey collies, we proposed that the crude colony-stimulating factor we had measured was the natural regulator for producing neutrophils [14]. Thus, we felt we were confirming Mackey and Glass' concept that cycling was due to an inherent delay in the natural feedback mechanism; the regulatory hormones were not really turned on until the neutrophil reserves were almost exhausted. Mike and I corresponded throughout this period, but we rarely met. We sent him data to analyze and criticize.

A major breakthrough for the cyclic neutropenia patients came with the cloning of genes for the colony-stimulating factors, and the biotechnology company, Amgen, producing enough of these glycoproteins for pre-clinical and clinical studies. Our first human studies were the treatment of cyclic neutropenia with the granulocyte colony-stimulating factor (G-CSF) product called "Neupogen." In 1986 I approached Amgen when it was really just a start-up company with the idea that a few weeks of G-CSF might raise the patients' counts enough to make them healthier and might stop neutrophil cycling. Of course, I did not know what would really happen. We found that daily injections of G-CSF shortened the period of severe neutropenia by shortening the neutrophil cycle from 21 to 14 days and almost completely prevented infections [16]. For the patients it was an overwhelming success! We hypothesized the shortening of the cycle was due to increased and accelerated production and more rapid delivery of neutrophils from the bone marrow to the blood. We later demonstrated these effects in normal human subjects [17]. Mike recently wrote to me reminding me that he had independently predicted these changes based upon his modeling work. He wrote that soon after I received his letter, I had replied with a preprint of our paper demonstrating these very effects.

Cyclic neutropenia is a rare disease first recognized more than a century ago. The diagnosis had always depended on a series of blood neutrophil counts over a long enough period of time to observe the regular cycles. Affected persons have regular signs and symptoms when the neutrophil counts are low, e.g., mouth ulcers, fevers, fatigue and "flu-like" symptoms, and sometimes very severe and life-threating infections. But sometimes it is not easy to distinguish the cycles because the levels are always quite low in some patients and periodicity of neutrophil cycles is not always apparent. There are also some patients who appear to have cyclic neutrophil oscillations and symptoms, but the more information you have the more uncertain you are about the diagnosis and whether the patient has another disease called "severe congenital neutropenia." One of the biggest problems for researchers and clinicians is that many patients simply have trouble getting a long series of blood cell counts to make a secure diagnosis. We discussed this problem with Mike, and he and his colleagues at McGill University developed a computer program that clinicians could use to determine if patients had cycles of their blood neutrophil counts [18].

At the same time, we were treating our first patients with G-CSF, researchers at Sloan-Kettering Cancer Center in New York showed that G-CSF was effective for treatment of severe congenital neutropenia and a few patients with neutropenia of unknown cause, a condition called "idiopathic neutropenia". With this information, we were confident enough to conduct a randomized clinical trial to prove the effectiveness of G-CSF for various types of severe chronic neutropenia, including cyclic neutropenia, under the sponsorship of Amgen. We did the trial and it was successful, changing the lives of many patients around the world [19]. It also provided new basic research opportunities and new opportunities to collaborate with Mike and his research team.

In the late 1980s there was a flurry of interest in using genetic sequencing to discover the causes for common and rare human diseases. The human genome project was just starting but methods were sufficient to begin the search, e.g., the search for the cystic fibrosis gene and many others. In 1949 Herbert Reiman wrote an excellent paper describing cyclic neutropenia as an autosomal dominant disease [20]. When we had a treatment for congenital and cyclic neutropenia, we began to discover more patients and families with this disorder. I thought we could find the genetic and molecular cause for cyclic neutropenia, i.e., the factor that Mike predicted to interrupt neutrophil production in his 1978 modeling paper, if we could just find enough patients and families. I was overly optimistic. It took 10 years: finding families, collecting clinical information and blood samples, countless hours of tedious laboratory work by a dedicated team of collaborators and better DNA sequencing methodologies to discover that a mutation in one gene, the gene for neutrophil elastase now called *ELANE*, causes cyclic neutropenia [21]. This gene serves as a backbone for production of a very potent protease synthesized at an early stage in neutrophil development. On September 23, 1999, Mike wrote to me, "I am delighted by the recent news of the identification of the mutation causing autosomal dominant cyclic neutropenia. I like your idea of relating this abnormality of a specific granule to accelerated apoptosis of early hematopoietic cells" [personal communication]. I appreciated that this finding fitted with Mike's concepts of how cycling might occur in his 1978 paper.

After finding that mutant *ELANE* caused cyclic neutropenia, it took many more years of work to show that the mutant protein triggers the death of developing neutrophils by the unfolded protein response and apoptosis [22]. We also discovered that some of our confusion about distinguishing severe cyclic neutropenia from severe congenital neutropenia is because both diseases are caused by mutations in *ELANE*, but there are many, more than 200, different *ELANE* mutations associated with cyclic and congenital neutropenia [23, 24]. There are a few overlapping mutations, mutations associated with both diseases. Through genotype-phenotype studies we discovered that there are mild and severe mutations, reflected by the dose of G-CSF to increase blood neutrophils and the risk of leukemic transformation [24]. As Mike predicted in 1978, some mutations cause severe disease with no apparent cycling, presumably because of very severe suppression of cells advancing beyond the promyelocyte stage, the stage in development when the neutrophil elastase protein is templated from the mutant gene. We believe now that with less severe mutations, cycling occurs because of the time required for myeloid recovery and the intrinsic delay in the response to neutropenia by the master endogenous regulator of neutrophil production, G-CSF.

About 20 years ago, I began to work with several young investigators in Mike's research group at McGill University. They were interested in both the methods for analyzing data to look for periodicity and looking for periodicity in other hematological diseases. It was fruitful work. From my perspective it was delightful to work with such bright young researchers who worked so hard to teach me advanced mathematics. It was also important because it strengthened our understanding that cyclic biological phenomena can be a consequence of diverse underlying physiological and pathological processes. The difference between the diseases we call severe cyclic neutropenia and severe congenital neutropenia (originally called "Kostmann's syndrome") are primarily due to variations in the mutations of the *ELANE* gene.

In 2008, I nominated Michael Mackey to return to Seattle as the University of Washington's Walker-Ames Lecturer. Members of the Departments of Medicine, Physiology, and Biophysics and Mathematics joined me for this invitation. In my letter to the chair of the selection committee, I wrote about the significance of his book, co-authored with Leo Glass, "From Clocks to Chaos: The Rhythms of Life." Mike was selected. He came and gave a superb lecture on "Bifurcations at the Bedside: How Non-linear Dynamics Can Help to Understand Periodic and Dynamical Diseases" [25]. His friends from his graduate school days, as well as many others, enjoyed being with him again.

In 2013 I participated in a symposium at the University of León to honor Mike on his 70th birthday. I believe I was the only non-mathematician attendee. We were all friends of Mike's, friends from all around the world. The talks were diverse and interesting. Equally impressive was the collegiality and appreciation for Michael Mackey's work. It is now 10 years later, and we are both less active but not retired or retiring. It has been so meaningful to me to have this long-term friendship.

References

1. Mackey, M.C.: The story of the 'Mackey-Glass' equation. In: Conte, G., Malnar, T. (eds.) Talk presented at: 17th IFAC workshop on time delay systems TDS 2.2. International Federation of Automatic Control, Montreal, CA (2022)
2. Mackey, M.C., Combs, J.W.: Tissue growth and homeostasis: consequences of control in synchronous cell populations. Growth. **38**(4), 477–494 (1974)
3. Burns, F.J., Tannock, I.F.: On the existence of a G 0 -phase in the cell cycle. Cell Tissue Kinet. **3**(4), 321–334 (1970). https://doi.org/10.1111/j.1365-2184.1970.tb00340.x. PMID: 5523059
4. Lund, J.E., Padgett, G.A., Ott, R.L.: Cyclic neutropenia in grey collie dogs. Blood. **29**(4), 452–461 (1967) PMID: 6067150

5. Dale, D.C., Alling, D.W., Wolff, S.M.: Cyclic hematopoiesis: the mechanism of cyclic neutropenia in grey collie dogs. J. Clin. Invest. **51**(8), 2197–2204 (1972). https://doi.org/ 10.1172/JCI107027. PMID: 5054472; PMCID: PMC292377

6. Yanay, O., Dale, D.C., Osborne, W.R.: Repeated lentivirus-mediated granulocyte colony-stimulating factor administration to treat canine cyclic neutropenia. Hum. Gene. Ther. **23**(11), 1136–1143 (2012). https://doi.org/10.1089/hum.2012.045. Epub 2012 Sep 12. PMID: 22845776; PMCID: PMC3498882

7. Guerry 4th, D., Dale, D.C., Omine, M., Perry, S., Wolff, S.M.: Periodic hematopoiesis in human cyclic neutropenia. J. Clin. Invest. **52**(12), 3220–3230 (1973). https://doi.org/10.1172/ JCI107522. PMID: 4750451; PMCID: PMC302598

8. Dale, D.C., Ward, S.B., Kimball, H.R., Wolff, S.M.: Studies of neutrophil production and turnover in grey collie dogs with cyclic neutropenia. J. Clin. Invest. **51**, 2190–2196 (1972) PMC292376

9. Cronkite EP, Fliedner TM. Granulocytopoiesis. N Engl. J. Med.. 1964 Jun 18 and 1964 June 25; 270: 1347–52 and 270: 1403–8. https://doi.org/10.1056/NEJM196406182702506 and https:// doi.org/10.1056/NEJM196406252702608. PMID: 14140268 and PMID: 14152874.

10. Craddock, C.G., Nakai, G.S.: Leukemic cell proliferation as determined by in vitro deoxyribonucleic acid synthesis. J. Clin. Invest. **41**(2), 360–369 (1962). https://doi.org/10.1172/ JCI104490. PMID: 13881943; PMCID: PMC289234

11. Athens, J.W., Haab, O.P., Raab, S.O., Mauer, A.M., Ashenbrucker, H., Cartwright, G.E., Wintrobe, M.M.: Leukokinetic studies. IV. The total blood, circulating and marginal granulocyte pools and the granulocyte turnover rate in normal subjects. J. Clin. Invest. **40**(6), 989–995 (1961). https://doi.org/10.1172/JCI104338. PMID: 13684958; PMCID: PMC290816

12. Dale, D.C., Graw Jr., R.G.: Transplantation of allogenic bone marrow in canine cyclic neutropenia. Science. **183**(4120), 83–84 (1974). https://doi.org/10.1126/science.183.4120.83. PMID: 4587264

13. Mackey, M.C.: Unified hypothesis for the origin of aplastic anamia and periodic hematopoiesis. Blood. **51**, 941–956 (1978)

14. Dale, D.C., Brown, C.H., Carbone, P., Wolff, S.M.: Cyclic urinary leukopoietic activity in grey collie dogs. Science. **173**(3992), 152–153 (1971). https://doi.org/10.1126/ science.173.3992.152. PMID: 5581910

15. Guerry 4th, D., Adamson, J.W., Dale, D.C., Wolff, S.M.: Human cyclic neutropenia: urinary colony-stimulating factor and erythropoietin levels. Blood. **44**(2), 257–262 (1974) PMID: 4852306

16. Hammond, W.P., Price, T.H., Souza, L.M., Dale, D.C.: Treatment of cyclic neutropenia with granulocyte colony stimulating factor. N. Engl. J. Med. **320**, 1306–1311 (1989)

17. Price, T.H., Gurkamal, S., Chatta, G.S., Dale, D.C.: The effect of recombinant granulocyte colony-stimulating factor on neutrophil kinetics in normal young and elderly humans. Blood. **88**, 335–340 (1996)

18. Dobbins, N.J., Bolyard, A.A., Chang, R.T., Self, J., Provencher Langlois, G., Mackey, M.C., Dale, D.C.: Application of spectral density/periodogram analysis to serial neutrophil counts to diagnose cyclic neutropenia. (ASH annual meeting abstracts). Blood. **126**, 4608 (2015)

19. Dale, D.C., Bonilla, M.A., Davis, M.W., Nakanishi, A., Hammond, W.P., Kurtzberg, J., Wang, W., Jakubowski, A., Winton, E., Lalezari, P., Robinson, W., Glaspy, J.A., Emerson, S., Gabrilove, J., Vincent, M., Boxer, L.A.: A randomized controlled phase III trial of recombinant human G-CSF for treatment of severe chronic neutropenia. Blood. **81**, 2496–2502 (1993) PMC4120868

20. Reimann, H.A., DeBerardinis, C.T.: Periodic (cyclic) neutropenia, an entity; a collection of 16 cases. Blood. **4**(10), 1109–1116 (1949) PMID: 18139383

21. Horwitz, M., Benson, K.F., Person, R.E., Aprikyan, A.G., Dale, D.C.: Mutations in ELA2, encoding neutrophil elastase, define a 21-day biological clock in cyclic haematopoiesis. Nat. Genet. **23**, 433–436 (1999) PMID: 10581030

22. Grenda, D.S., Murakami, M., Ghatak, J., Xia, J., Boxer, L.A., Dale, D., Dinauer, M.C., Link, D.C.: Mutations of the ELA2 gene found in patients with severe congenital neutropenia induce the unfolded protein response and cellular apoptosis. Blood. **110**, 4179–4187 (2007) PMC2234798

23. Dale, D.C., Person, R.E., Bolyard, A.A., Aprikyan, A.G., Bos, C., Bonilla, M.A., Boxer, L.A., Kannourakis, G., Zeidler, C., Welte, K., Benson, K.F., Horwitz, M.: Mutations in the gene encoding neutrophil elastase in congenital and cyclic neutropenia. Blood. **96**(7), 2317–2322 (2000)

24. Makaryan, V., Zeidler, C., Bolyard, A.A., Skokowa, J., Rodger, E., Kelley, M.L., Boxer, L.A., Bonilla, M.A., Newburger, P.E., Shimamura, A., Zhu, B., Rosenberg, P.S., Link, D.C., Welte, K., Dale, D.C.: The diversity of mutations and clinical outcomes for *ELANE* associated neutropenia. Curr. Op. Hematol. **22**, 3–11 (2015) PMC4380169

25. Mackey, M.C. (Centre for Applied Mathematics in Bioscience and Medicine, Dept of Physiology, McGill University, Quebec, CA): Bifurcations at the bedside: how non-linear dynamics can help to understand periodic and dynamical disease [Lecture notes]. Lecture presented at: Walker-Ames Lecture Series: Michael C. Mackey (The Graduate School, University of Washington, Seattle, WA). 2009 Apr 09

Michael Mackey and Data-Driven Models of Chemotherapy-Induced Neutropenia

Morgan Craig

Abstract During his research career, Michael Mackey, Joseph Morley Drake Professor Emeritus of Physiology in the Department of Physiology at McGill University, Canada, made numerous consequential insights into biological mechanisms of dynamic hematologic diseases and gene regulatory networks. Throughout, he emphasized the importance of data on mathematical model conceptualization, development, calibration, and validation. This brief review describes contributions by Michael Mackey and his collaborators toward understanding and treating chemotherapy-induced neutropenia, a frequent and serious toxic side effect of cytotoxic chemotherapy, using data-driven mathematical models.

1 Introduction

The human body maintains homeostasis through multiple overlapping networks, implying that physiological data are inherently multiscale and multidimensional [1, 2]. Accordingly, revealing mechanistic drivers of physiological responses requires a multipronged approach integrating mathematical and computational modelling [3]. Historically, one of the greatest difficulties faced by mathematical modellers in biomedicine is a lack of data. It was not until recent advances in data collection and analysis that mathematical models in physiology could become strongly anchored within experimental and clinical paradigms.

During his career, Michael Mackey worked on a variety of problems ranging from the abstract [4–7] to those with concrete applications for patient care [8–11]. Throughout, he placed a special emphasis on *"biologically realistic mathematical models, careful consideration of laboratory and/or clinical data, and achieving*

M. Craig (✉)
Sainte-Justine University Hospital Research Centre, Montréal, QC, Canada

Department of Mathematics and Statistics, Université de Montréal, Montréal, QC, Canada
e-mail: morgan.craig@umontreal.ca

© The Author(s), under exclusive license to Springer Nature Switzerland AG 2025   9
Y. Mori et al. (eds.), *Dynamics of Physiological Control*, Lecture Notes
on Mathematical Modelling in the Life Sciences,
https://doi.org/10.1007/978-3-031-82396-1_2

a reasonable concordance between the data and the modelling," [12] despite carrying out his early research at a time during which bulk data were more difficult to obtain. One focal point in later work was chemotherapy-induced neutropenia (i.e., a lack of neutrophils), which is one of the most frequent toxic side-effects of cytotoxic chemotherapy. Neutrophils are innate immune cells that have been historically thought to respond to bacterial and fungal infections [13], though increasing evidence points to their importance of in viral infections [13, 14]. Neutropenia leaves individuals open to opportunistic infections [15] and can therefore be particularly dangerous to a cancer patient. Chemotherapy-induced neutropenia requires treatment adaptation and/or cessation [15, 16]. Hence, there is an important need for methods that can predict neutropenic periods, potentially staving off complete treatment discontinuance [17].

Multiple authors have previously reviewed mathematical models in medicine/ physiology [18] from the lens of e.g., cancer growth and treatment [19], kidney function [20], or a general overview of the applications of models to medical applications [21–23]. This brief review will focus on Michael Mackey's work modelling chemotherapy-induced neutropenia and the integration of these mechanistic mathematical models with experimental and clinical data to advance patient care.

2   Origins: Cyclic Neutropenia and the G0 Model of Hematopoietic Stem Cell Division

In blood cell modelling, Michael Mackey initially focused on cyclic neutropenia, a rare genetic condition during which neutrophil counts oscillate at regular frequencies throughout life. This work was particularly influenced by his collaboration with David C. Dale (see previous review [11] and article in this collection). To study cyclic neutropenia using mathematical modelling, Michael Mackey and his co-authors combined the G0 mathematical model of hematopoietic stem cell (HSC) cycling and quiescence [24, 25] with delay differential equation models describing the development of terminally differentiated neutrophils in the blood [26–31] (Fig. 1a). The combined model included HSCs, and neutrophil progenitors that underwent exponential expansion and a period of maturation before egressing out of the bone marrow into circulation.

Perhaps driven by the need to accurately parameterize these models, in the early 2000s, Michael Mackey and colleagues published several articles defining key kinetic rates in HSCs. These include a 2001 paper [33] in which Michael Mackey calculated the rates of proliferation, apoptosis, and self-renewal from mouse and cat data to infer the number of effective cell divisions per HSC lifetime and their net amplification. In 2003, Mackey, Aprikyan, and Dale integrated data of CD15+ cells (i.e., post-mitotic neutrophil progenitors) from 12 individuals to estimate the rate of neutrophil progenitor apoptosis and thus the granulocyte turnover rate [34]. From their estimate of the apoptotic cell death rate, they calculated the normal

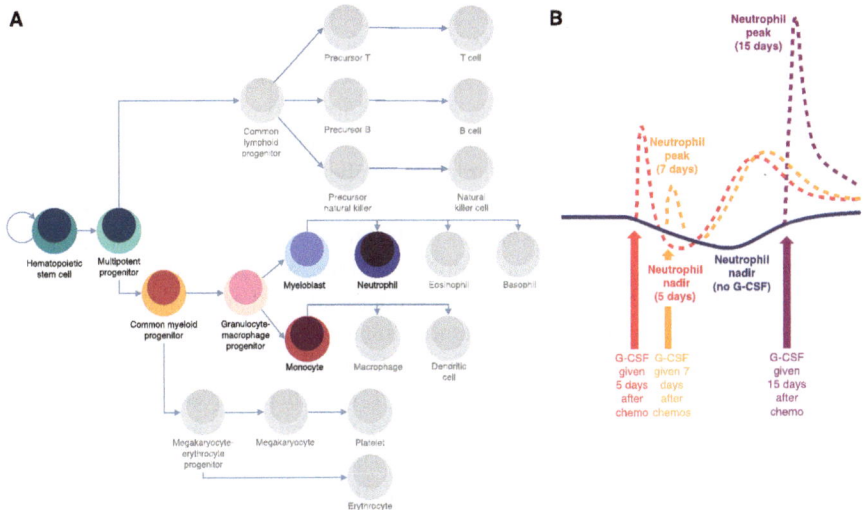

Fig. 1 Neutrophil and monocyte production, and the impact of the administration of granulocyte colony-stimulating factor agents on chemotherapy-induced neutropenia. (**a**) The process of hematopoiesis, from hematopoietic stem cell to terminally differentiated blood cells. Highlighted in colour are neutrophils and monocytes. Adapted from Zhang et al. [32] (**b**) The timing of exogenous G-CSF agents determines the degree of neutropenia and neutrophilia

half-life of these cells to be around 10.5 hours. This short neutrophil half-life (or half-removal time [35]) explains why neutrophils are particularly affected by cytotoxic chemotherapy, given that they are under constant production within the bone marrow to replenish the circulating neutrophil pool.

3 Early Models of Chemotherapy-Induced Neutropenia

Chemotherapy-induced neutropenia is treated using granulocyte colony-stimulating factor (G-CSF) agents [14]. G-CSF is the main cytokine responsible for neutrophil production and acts in negative feedback with neutrophil counts in the blood to ensure their homeostatic levels [36]. Chemotherapy-driven effects on neutrophil counts disrupt this feedback system. G-CSF agents are administered to rescue neutrophils and reduce the duration of neutropenia. However, there is a lack of clinical consistency as to the timing of G-CSF relative to the onset of neutropenia [37]. Using mechanistic mathematical models of neutrophil production, several groups [38–40], including our own [36, 41], have found the timing of G-CSF administration to be determined by the neutrophil dynamics themselves. For example, treating too late, once neutrophil counts are near their nadir, or too early, before neutrophil counts have begun decreasing, can worsen the degree or duration of neutropenia (Fig. 1b).

In 2006, Michael Mackey and his group began focusing on the administration of exogeneous G-CSF in the context of cyclical neutropenia. Foley, Bernard, and Mackey [42] studied different treatment schedules to attenuate neutrophil cycling using data from grey collies, a breed known to experience canine cyclic neutropenia. The four studied protocols were 1) a daily, phase-dependent schedule determined through theoretical bifurcation analyses (current cyclic neutropenia treatment protocols in humans administer G-CSF daily), 2) every second day, 3) administration of G-CSF based on the neutrophil concentration dipping below a predetermined threshold, and 4) a random schedule. With a parameterized model, Foley et al. showed that protocols 2 and 3 reduced the total amount of administered G-CSF thus lessening the patient's potential financial costs and treatment burden.

By extending the neutrophil production and G-CSF model to include chemotherapeutic effects, Zhuge, Lei, and Mackey [43] studied the effects of periodic chemotherapy and G-CSF without explicitly modelling the pharmacokinetics and pharmacodynamics (PK/PD) of either drug. Their results predicted that periodic chemotherapy may induce the phenomenon of resonance (i.e., large amplitude oscillations matched to the system's natural vibration period) in the neutrophil population, which could aggravate neutropenia. They also predicted highly variable responses to G-CSF administration with respect to the timing of G-CSF within each chemotherapy period.

4   Integrating Drug Pharmacokinetics: The Importance of Scheduling for Supportive G-CSF Therapy During Chemotherapy

Brooks et al. [44] then expanded the work of Zhuge et al. [43] by integrating pharmacokinetic models for both drugs. For the chemotherapeutic agent, they used an exponential decay model and constructed a two-compartment PK model modelling the exchange of G-CSF between the tissues and the plasma to describe changes to plasma G-CSF concentrations. Through model simulations, they showed that the timing between chemotherapy administrations could induce resonance that could be counteracted by early administrations of G-CSF, but the resonance would be amplified by G-CSF administered too late. Further, their model predicted that the neutrophil nadir was driven primarily by the effects of chemotherapy and not G-CSF.

Brooks et al. did not include comparisons of their model's predictions to clinical data. In two earlier studies, Pfreundschuh et al. described the results of clinical trials comparing CHOP (cyclophosphamide, doxorubicin, vincristine, and prednisone) and CHOEP (CHOP plus etoposide) with supportive G-CSF to treat younger [45, 46] and older adults [43] with aggressive lymphomas. Based on their results, in work with Michael Mackey, we integrated a population PK model of a chemotherapeutic agent [47] and simulated the average response to treatment with 14-day periodic

CHOP with G-CSF support on days 4–13 (total of 10 doses per chemotherapy cycle, or 60 doses per treatment per patient) [36]. Our model showed good agreement to the Pfreundschuh et al. clinical trial results [43], without calibrating to these data. We then explored the optimal timing of both chemotherapy and subcutaneous G-CSF. Our results showed that administering G-CSF 6–7 days post-chemotherapy meaningfully reduced the amount of G-CSF needed within each chemotherapy cycle to just 3–4 per cycle (i.e., total of 18–24 per treatment per patient), reducing patient drug burden.

Interestingly, the 2015 Craig et al. [36] model identified an important issue with empirically based models of G-CSF pharmacokinetics, namely that they overestimated the contribution of renal elimination to G-CSF pharmacokinetics. Like all cytokines, G-CSF binds to receptors on the surface of target cells before being internalized, and upregulating gene products to induce its effect. At basal concentrations, this binding is the dominant mode of elimination, with linear (renal) clearance used as an emergency clearance mechanism when plasma concentrations rise too high. By establishing PK models for G-CSF from data following exogenous administration, previous models underestimated cell-mediated clearance relative to kidney elimination, thus hampering their use at physiological concentrations. We demonstrated this in Craig et al. 2016 [48] by varying the percent of renal elimination in our model and showing that too large a contribution caused unrealistic neutrophil/G-CSF dynamics (i.e., high G-CSF with high neutrophil concentrations. We further developed an explicit model of G-CSF binding and unbinding and used it to show that the hypothesis that cytokines are in quasi-equilibrium at homeostatic concentrations did not hold, which has implications for cytokines that are administered exogenously to treat various diseases (e.g., interferon (IFN) alpha [49, 50], IFN beta [46], and interleukin-2 [51]).

5   Leveraging Data to Identify Biomarkers of Chemotherapy-Induced Neutropenia

Though neutropenia clearly presents a health risk to patients receiving cytotoxic chemotherapy, neutrophilia (neutrophil counts above the reference range) can also be problematic in the cancer context. For example, the formation of neutrophil extracellular traps or NETs [52] has been shown to worsen cancer prognoses [52–56]. Thus, the judicious administration of G-CSF during cytotoxic chemotherapy is important, as G-CSF strongly stimulates neutrophil production. In Mackey et al. [57], we studied the question of how to simultaneously mitigate both neutropenia and neutrophilia by examining data from 286 children diagnoses with acute lymphoblastic leukemia (ALL) treated according to the Dana-Farber Cancer Institute (DFCI) ALL Consortium protocols (DFCI 87–01, DFCI 91–01, DFCI 95–01, or DFCI 00–01). In these patients, we determined that 26% presented neutrophil dynamics with resonance phenomena driven by cyclic (once-every-three-weeks) cytotoxic chemotherapy. Leveraging the Craig et al. [48] model, we then used simulations to determine which G-CSF schedules post-chemotherapy best reduced neutropenic and neutrophilic episodes, thereby reducing the possibility of resonance behaviour. We found that G-CSF administered 6–7 days after chemotherapy, as neutrophil concentrations began to drop, would limit neutropenia and neutrophilia, similar to predictions from previous models [36, 39, 58]. Thus, through the tight integration with clinical data emphasized throughout Michael Mackey's career, this study helped to establish treatment schedules with potentially high clinically relevance.

Finally, we further expanded the neutrophil production model by including the production of monocytes in the bone marrow (monocytopoiesis) [59]. Like neutrophils, monocytes are granulocytes, however once out of the bone marrow, monocytes differentiate into dendritic cells that act to present antigens to T lymphocytes in lymph nodes or macrophages that perform important phagocytotic functions during infections (Fig. 1a). From the same pediatric patient data analyzed in Michael Mackey et al. [57], we identified periods of monocytopenia that were correlated with neutropenic episodes [59]. Through model simulations, our predictions suggested that monocytopenia tended to precede neutropenia, owing to the time required to produce each cell in the bone marrow and the blood. Our results therefore suggest that falling monocyte concentrations are a potential biomarker for neutropenic episodes, and thus a signal to initiate supportive G-CSF therapy during cytotoxic anti-cancer treatment, again highlighting the importance of data integration into mechanistic mathematical models to provide clinically valuable insights.

6 Conclusion

The complexity of human physiology demands that we develop complementary methods to best understand it [18]. Mathematical and computational modelling has proven critical for identifying regulatory mechanisms and key dynamics that may otherwise be difficult or impossible to study using experimental or clinical approaches. That said, models built without incorporating data may be limited in their reach and are restricted in their ability to answer specific physiological questions. In his work on chemotherapy-induced neutropenia, Michael Mackey demonstrated how mathematical modelling can help guide clinical practices through the incorporation of physiological mechanisms and animal/human data. Together with many other researchers in mathematical biology, his work thus laid the framework for modern mechanistic modelling approaches in medicine, with important implications for meaningfully advancing biomedical research.

Acknowledgements I would like to thank Michael Mackey for years of supervision and guidance. I am particularly thankful to our collaborators for their contributions to work cited within this review and beyond.

Conflicts of Interest The author has no conflicts of interest to declare.

Funding

MC was funded by a J1 Fonds de recherche du Québec-Santé (FRQS) Research Scholar award.

References

1. Liang, J.: A review of multiscale science: Materials, biology, multiscale data analysis and examples from complex physiological systems, pp. 1360–1365. IEEE (2017)
2. Groen, D., Zasada, S.J., Coveney, P.V.: Survey of multiscale and multiphysics applications and communities. Comput. Sci. Eng. **16**(2), 34–43 (2014). https://doi.org/10.1109/MCSE.2013.47
3. Araujo, R.P., McElwain, D.L.S.: A history of the study of solid tumour growth: the contribution of mathematical modelling. Bull. Math. Biol. **66**(5), 1039–1091 (2004). https://doi.org/10.1016/j.bulm.2003.11.002
4. Mackey, M.C., Tyran-Kamińska, M.: How can we describe density evolution under delayed dynamics? Chaos. **31**(4), 043114 (2021). https://doi.org/10.1063/5.0038310
5. Losson, J., Mackey, M.C., Taylor, R., Tyran-Kamińska, M.: Density evolution under delayed dynamics Fields Institute Monographs. Springer, New York, NY (2020)
6. Self, J., Mackey, M.C.: Random numbers from a delay equation. J. Nonlinear Sci. **26**(5), 1311–1327 (2016). https://doi.org/10.1007/s00332-016-9306-9
7. Lei, J., Mackey, M.C.: Deterministic Brownian motion generated from differential delay equations. Phys. Rev. E. **84**(4), 041105 (2011). https://doi.org/10.1103/PhysRevE.84.041105
8. Milton, J.G., Mackey, M.C.: Periodic haematological diseases: mystical entities or dynamical disorders? J. R. Coll. Physicians Lond. **23**(4), 236–241 (1989)
9. Steward, C.G., Groves, S.J., Taylor, C.T., et al.: Neutropenia in Barth syndrome: characteristics, risks, and management. Curr. Opin. Hematol. **26**(1), 6–15 (2019). https://doi.org/10.1097/moh.0000000000000472
10. Lei, J., Mackey, M.C.: Understanding and treating cytopenia through mathematical modeling. In: Corey, S.J., Kimmel, M., Leonard, J.N. (eds.) A systems biology approach to blood, pp. 279–302: Chapter 14. Springer, New York (2014)
11. Dale, D.C., Mackey, M.C.: Understanding, treating and avoiding hematological disease: better medicine through mathematics? Bull. Math. Biol. **77**(5), 739–757 (2014). https://doi.org/10.1007/s11538-014-9995-x
12. Mackey, M.C., Michael, C.: Mackey - Emeritus Professor. Department of Physiology - McGill University (2023) Accessed December 18, 2023, https://www.mcgill.ca/physiology/directory/core-faculty/michael-mackey
13. Galani, I.E., Andreakos, E.: Neutrophils in viral infections: current concepts and caveats. J. Leukoc. Biol. **98**(4), 557–564 (2015). https://doi.org/10.1189/jlb.4VMR1114-555R
14. Zhang, Y., Wang, Q., Mackay, C.R., Ng, L.G., Kwok, I.: Neutrophil subsets and their differential roles in viral respiratory diseases. J. Leukoc. Biol. **111**(6), 1159–1173 (2022). https://doi.org/10.1002/jlb.1mr1221-345r
15. Blayney, D.W., Schwartzberg, L.: Chemotherapy-induced neutropenia and emerging agents for prevention and treatment: a review. Cancer Treat. Rev. **109**, 102427 (2022). https://doi.org/10.1016/j.ctrv.2022.102427
16. Boccia, R., Glaspy, J., Crawford, J., Aapro, M.: Chemotherapy-induced neutropenia and febrile neutropenia in the US: a beast of burden that needs to be tamed? Oncologist. **27**(8), 625–636 (2022). https://doi.org/10.1093/oncolo/oyac074

17. Craig, M.: Towards quantitative systems pharmacology models of chemotherapy-induced neutropenia. CPT: Pharmacomet. Syst. Pharmacol. **6**(5), 293–304 (2017). https://doi.org/10.1002/psp4.12191
18. Keener, J., Sneyd, J.: Mathematical physiology Interdisciplinary applied mathematics. Springer, New York, NY (2009)
19. Chaplain, M.A.J.: Multiscale mathematical modelling in biology and medicine. IMA J. Appl. Math. **76**(3), 371–388 (2011). https://doi.org/10.1093/imamat/hxr025
20. Pack, A.I., Murray-Smith, D.J.: Mathematical models and their applications in medicine. Scott. Med. J. **17**(12), 401–409 (1972). https://doi.org/10.1177/003693307201701205
21. Chambers, R.B.: The role of mathematical modeling in medical research: "research without patients?". Oschner J. **2**(4), 218–223 (2000)
22. Liu, Y., Wu, R., Yang, A.: Research on medical problems based on mathematical models. Mathematics. **11**(13), 2842 (2023)
23. Smye, S.W., Clayton, R.H.: Mathematical modelling for the new millenium: medicine by numbers. Med. Eng. Phys. **24**(9), 565–574 (2002). https://doi.org/10.1016/S1350-4533(02)00049-8
24. Pujo-Menjouet, L., Bernard, S., Mackey, M.C.: Long period oscillations in a G0 model of hematopoietic stem cells. SIAM J. Appl. Dyn. Syst. **4**(2), 312–332 (2005). https://doi.org/10.1137/030600473
25. Brunetti, M., Mackey, M.C., Craig, M.: Understanding normal and pathological hematopoietic stem cell biology using mathematical modelling. Curr. Stem Cell Rep. **7**, 109–120 (2021). https://doi.org/10.1007/s40778-021-00191-9
26. Hearn, T., Haurie, C., Mackey, M.C.: Cyclical neutropenia and the peripherial control of white blood cell production. J. Theor. Biol. **192**, 167–181 (1998)
27. Haurie, C., Dale, D.C., Mackey, M.C.: Occurrence of periodic oscillations in the differential blood counts of congenital, idiopathic, and cyclical neutropenic patients before and during treatment with G-CSF. Exp. Hematol. **27**, 401–409 (1999). https://doi.org/10.1016/S0301-472X(98)00061-7
28. Bernard, S., Bélair, J., Mackey, M.C.: Oscillations in cyclical neutropenia: new evidence based on mathematical modeling. J. Theor. Biol. **223**, 1–16 (2015)
29. Haurie, C., Dale, D.C., Mackey, M.C., Haurie, B.C.: Cyclical neutropenia and other periodic hematological diseases: a review of mechanisms and mathematical models. Blood. **92**, 2629–2640 (1998)
30. Hearn, T., Haurie, C., Mackey, M.C.: Cyclical neutropenia and the peripheral control of white blood cell production. J. Theor. Biol. **192**(2), 167–181 (1998). https://doi.org/10.1006/jtbi.1997.0589
31. Colijn, C., Fowler, A.C., Mackey, M.C.: High frequency spikes in long period blood cell oscillations. J. Math. Biol. **53**(4), 499–519 (2006). https://doi.org/10.1007/s00285-006-0027-9
32. Zhang, P., Zhang, C., Li, J., Han, J., Liu, X., Yang, H.: The physical microenvironment of hematopoietic stem cells and its emerging roles in engineering applications. Stem Cell Res. Ther. **10**(1), 327 (2019). https://doi.org/10.1186/s13287-019-1422-7
33. Mackey, M.C.: Cell kinetic status of haematopoietic stem cells. Cell Prolif. **34**(2), 71–83 (2001). https://doi.org/10.1046/j.1365-2184.2001.00195.x
34. Mackey, M.C., Aprikyan, A.A.G., Dale, D.C.: The rate of apoptosis in post mitotic neutrophil precursors of normal and neutropenic humans. Cell Prolif. **36**(1), 27–34 (2003). https://doi.org/10.1046/j.1365-2184.2003.00251.x
35. Craig, M., Humphries, A.R., Mackey, M.C.: An upper bound for the half-removal time of neutrophils from circulation. Blood. **128**(15), 1989–1991 (2016). https://doi.org/10.1182/blood-2016-07-730325
36. Craig, M., Humphries, A.R., Nekka, F., et al.: Neutrophil dynamics during concurrent chemotherapy and G-CSF administration: mathematical modelling guides dose optimisation to minimise neutropenia. J. Theor. Biol. **385**, 77–89 (2015). https://doi.org/10.1016/j.jtbi.2015.08.015

37. Hanna, K.S., Mancini, R., Wilson, D., Zuckerman, D.: Comparing granulocyte Colony-stimulating factors prescribing practices versus guideline recommendations in a large community cancer center. J. Hematol. Oncol. Pharm. **9**(3), 121–126 (2019)

38. Scholz, M., Engel, C., Loeffler, M.: Modelling human granulopoiesis under polychemotherapy with G-CSF support. J. Math. Biol. **50**, 397–439 (2005). https://doi.org/10.1007/s00285-004-0295-1

39. Vainas, O., Ariad, S., Amir, O., et al.: Personalising docetaxel and G-CSF schedules in cancer patients by a clinically validated computational model. Br. J. Cancer. **107**, 814–822 (2012). https://doi.org/10.1038/bjc.2012.316

40. Yankelevich, M., Hoogstra, D.J., Abrams, J., Chu, R., Bhambhani, K., Taub, J.W.: Delayed granulocyte colony-stimulating factor (G-CSF) administration after chemotherapy reduces total G-CSF doses without affecting neutrophil recovery in a randomized clinical study in children with solid tumors. Pediatr. Hematol. Oncol. **37**(8), 665–675 (2020). https://doi.org/10.1080/08880018.2020.1779885

41. Paredes Bonilla, R.V., Nekka, F., Craig, M.: A quantitative systems pharmacology framework for optimal doxorubicin granulocyte colony-stimulating factor regimens in triple-negative breast cancer. Pharmacology. **106**(9–10), 542–550 (2021). https://doi.org/10.1159/000518037

42. Foley, C., Bernard, S., Mackey, M.C.: Cost-effective G-CSF therapy strategies for cyclical neutropenia: mathematical modelling based hypotheses. J. Theor. Biol. **238**(4), 756–763 (2006)

43. Zhuge, C., Lei, J., Mackey, M.C.: Neutrophil dynamics in response to chemotherapy and G-CSF. J. Theor. Biol. **293**, 111–120 (2012)

44. Brooks, G., Provencher, G., Lei, J., et al.: Neutrophil dynamics after chemotherapy and G-CSF: the role of pharmacokinetics in shaping the response. J. Theor. Biol. **315**, 97–109 (2012). https://doi.org/10.1016/j.jtbi.2012.08.028

45. Pfreundschuh, M., Trümper, L., Kloess, M., et al.: Two-weekly or 3-weekly CHOP chemotherapy with or without etoposide for the treatment of young patients with good-prognosis (normal LDH) aggressive lymphomas: results of the NHL-B1 trial of the DSHNHL. Blood. **104**(3), 626–633 (2004). https://doi.org/10.1182/blood-2003-06-2094

46. Pfreundschuh, M., Trümper, L., Kloess, M., et al.: Two-weekly or 3-weekly CHOP chemotherapy with our without etoposide for the treatment of elderly patients with aggressive lymphomas: results of the NHL-B2 trial of the DSHNHL. Blood. **104**, 634–641 (2004)

47. Pérez-Ruixo, C., Valenzuela, B., Fernández Teruel, C., et al.: Population pharmacokinetics of PM00104 (Zalypsis(R)) in cancer patients. Cancer Chemother. Pharmacol. **69**, 15–24 (2012)

48. Craig, M., Humphries, A.R., Mackey, M.C.: A mathematical model of granulopoiesis incorporating the negative feedback dynamics and kinetics of G-CSF/neutrophil binding and internalisation. Bull. Math. Biol. **78**(12), 2304–2357 (2016). https://doi.org/10.1007/s11538-016-0179-8

49. Mocellin, S., Pasquali, S., Rossi, C.R., Nitti, D.: Interferon alpha adjuvant therapy in patients with high-risk melanoma: a systematic review and meta-analysis. JNCI J. Natl. Cancer Inst. **102**(7), 493–501 (2010). https://doi.org/10.1093/jnci/djq009

50. Cohen, J.A., Comi, G., Selmaj, K.W., et al.: Safety and efficacy of ozanimod versus interferon beta-1a in relapsing multiple sclerosis (RADIANCE): a multicentre, randomised, 24-month, phase 3 trial. Lancet Neurol. **18**(11), 1021–1033 (2019). https://doi.org/10.1016/S1474-4422(19)30238-8

51. Yuan, Y., Kolios, A.G.A., Liu, Y., et al.: Therapeutic potential of interleukin-2 in autoimmune diseases. Trends Mol. Med. **28**(7), 596–612 (2022). https://doi.org/10.1016/j.molmed.2022.04.010

52. Cools-Lartigue, J., Spicer, J., Najmeh, S., Ferri, L.: Neutrophil extracellular traps in cancer progression. Cell. Mol. Life Sci. **71**(21), 4179–4194 (2014). https://doi.org/10.1007/s00018-014-1683-3

53. Shinde-Jadhav, S., Mansure, J.J., Rayes, R.F., et al.: Role of neutrophil extracellular traps in radiation resistance of invasive bladder cancer. Nat. Commun. **12**(1), 2776 (2021). https://doi.org/10.1038/s41467-021-23086-z

54. Cools-Lartigue, J., Spicer, J., McDonald, B., et al.: Neutrophil extracellular traps sequester circulating tumor cells and promote metastasis. J. Clin. Invest. **123**(8), 3446–3458 (2013). https://doi.org/10.1172/JCI67484

55. Adrover, J.M., McDowell, S.A.C., He, X.-Y., Quail, D.F., Egeblad, M.: NETworking with cancer: the bidirectional interplay between cancer and neutrophil extracellular traps. Cancer Cell. **41**(3), 505–526 (2023). https://doi.org/10.1016/j.ccell.2023.02.001

56. Milette, S., De Meo, M., Mongeon, B., et al.: 76P circadian control of neutrophil extracellular trap formation temporally regulates metastatic lung cancer progression. ESMO Open. **8**(1), 101886 (2023). https://doi.org/10.1016/j.esmoop.2023.101886

57. Mackey, M.C., Glisovic, S., Leclerc, J.-M., Pastore, Y., Krajinovic, M., Craig, M.: The timing of cyclic cytotoxic chemotherapy can worsen neutropenia and neutrophilia. Br. J. Clin. Pharmacol. **87**, 687–693 (2020). https://doi.org/10.1111/bcp.14424

58. Scholz, M., Engel, C., Loeffler, M.: Model-based design of chemotherapeutic regimens that account for heterogeneity in leucopoenia. Br. J. Haematol. **132**, 723–735 (2006). https://doi.org/10.1111/j.1365-2141.2005.05957.x

59. Cassidy, T., Humphries, A.R., Craig, M., Mackey, M.C.: Characterizing chemotherapy-induced neutropenia and Monocytopenia through mathematical modelling. Bull. Math. Biol. **82**(8), 104–104 (2020). https://doi.org/10.1007/s11538-020-00777-0

Biological Rhythms: Mechanisms, Functions, and Associated Disorders

Albert Goldbeter

Abstract Oscillations abound in biological systems. Examples range from cardiac and neural rhythms to a variety of cellular rhythms of nonelectrical nature, such as glycolytic oscillations in yeast, cyclic AMP oscillations and waves in *Dictyostelium* amoebae, Ca^{++} oscillations, the segmentation clock responsible for somite formation in embryonic development, circadian clocks driving the sleep-wake cycle in mammals, and the oscillator driving the cell cycle. The mathematical modeling of these oscillatory phenomena shows how biological rhythms display common properties despite marked differences in their period and molecular mechanism. Physiological disorders in the form of dynamical diseases are often associated with the malfunction of these rhythms. Multiple layers of regulation by feedback loops and the cooperativity of biological processes provide the sources of nonlinearity responsible for the onset of oscillatory behavior in biological systems.

1 Introduction

Many of the most important physiological functions possess a rhythmic nature: respiration, operation of the heart, periodic contractions of the intestine, uterus contractions at the time of delivery, generation of brain rhythms, and the sleep-wake cycle controlled by the circadian clock. Besides these rhythms, which are immediately perceptible, many biological processes oscillate within our cells or

---

Dedication: It is a pleasure to dedicate this article to Michael C. Mackey, pioneer of Mathematical Physiology, long-time colleague and friend.

A. Goldbeter (✉)
Unit of Theoretical Chronobiology, Faculty of Sciences, Université Libre de Bruxelles (ULB), Brussels, Belgium
e-mail: albert.goldbeter@ulb.be

© The Author(s), under exclusive license to Springer Nature Switzerland AG 2025   19
Y. Mori et al. (eds.), *Dynamics of Physiological Control*, Lecture Notes
on Mathematical Modelling in the Life Sciences,
https://doi.org/10.1007/978-3-031-82396-1_3

body without our being aware of these rhythms, with periods ranging from a fraction of a second—as in nerve or muscle cells—to seconds, minutes, or more.

Oscillations also occur in some chemical systems, the prototype of which remains the Belousov-Zhabotinsky reaction, discovered some six decades ago [1–3]. Periodic behavior is also observed in electrochemical systems and in electrical circuits. Although periodic phenomena occur in physics and chemistry, one can only be struck by the ubiquitousness of oscillations in biological systems [4–10]. To understand why biological rhythms are so abundant, it is necessary to look at their functions and at their underlying mechanisms. This will lead us to look at conditions in which disturbances of the rhythms are associated with physiological disorders. Our goal here is to present lines of reflection without giving detailed references for all the examples of oscillatory behavior that will be mentioned (for additional references see Goldbeter [7–9]).

2 Functions of Biological Rhythms: From Repetitive Machines to the Frequency Encoding of Intercellular Signals

The fact that biological rhythms are so ubiquitous implies that they were retained in the course of evolution because they perform indispensable functions (see Table 1). This is particularly clear in the case of respiration and of the heart, which behave as repetitive machines. The respiratory rhythm is controlled by a neural network that gives rise, repetitively, to inspiration and expiration [11]. Each of these two phases is followed by the other so as to generate a spontaneous rhythm. In the case of the heart [12, 13], cardiac cells located in the sinus node spontaneously and periodically initiate the generation of an action potential that will propagate to the ventricles

Table 1 Some major physiological functions associated with biological rhythms [9]

Neuronal oscillations in cognitive or sensory processes
Neuronal control of movements: central pattern generators
Periodic contraction of the heart
Respiratory rhythm
Periodic intestinal contractions (generated by interstitial Cajal cells)
90-min cycles of alternating REM-nonREM phases in sleep
Sleep-wake cycle (controlled by the circadian clock)
Nutrition cycle (controlled by the circadian clock)
Cell division
Hormonal rhythms and pulsatile hormone secretion (e.g., GnRH, growth hormone, insulin)
Ovulation
Fertilization-induced Ca^{++} oscillations trigger the onset of egg development
Uterine contractions (triggered by Ca^{++} oscillations)
Seasonal rhythms (flowering, animal reproduction, migrations)

and cause their contraction. In both cases, the periodic nature of the physiological process is inherent to its function.

In a certain sense, the cell cycle can also be viewed as a repetitive machine. The biochemical network that underlies the ordered progression through the G1, S (DNA replication), G2 and M (mitosis) phases in mammalian cells spontaneously leads to the initiation of a new G1 phase after a cell divides [14]. A network of cyclin-dependent kinases (Cdks) controls the passage from one cell cycle phase to the next. The network is organized in such a way that the activation of a one cyclin-Cdk module leads to its inactivation and to the activation of the next module in the network [15].

Another major function of biological rhythms is their capability to allow for the frequency encoding of intracellular or intercellular signals. This is observed in the brain, where trains of action potentials of varying frequencies underlie intercellular communication involved in sensory responses and, ultimately, consciousness [4, 16]. Frequency coding is by no means restricted to electrical signaling. Hormonal signals are often generated in a pulsatile rather than continuous manner [17]. The frequency of the hormonal pulses governs their physiological function. A striking example if provided by the hormone GnRH, which elicits the secretion of the gonadotropic hormones LH and FSH by the pituitary. In mammals GnRH is secreted in a pulsatile manner by the hypothalamus with a frequency of the order of one 5-min pulse every hour [18].

In a series of remarkable experiments, Knobil and co-workers showed around 1980 that the frequency of the pulses is of key significance for their physiological efficiency [19, 20]. Pulses of GnRH that are either too frequent or too rare— or occur randomly, in a non-periodic manner—fail to elicit the release of LH and FSH at levels capable of inducing ovulation. Only the pulses delivered at the physiological frequency succeed to trigger the release of the gonadotropic hormones. These observations led to the restoration of ovulation in women lacking proper gonadotropic stimulation by LH and FSH due to abnormal patterns of GnRH secretion, after the latter hormone was delivered by a pump programmed to deliver GnRH at the physiological frequency [21].

Another example of pulsatile signaling in intercellular communication is provided by the aggregation of the cellular slime mold *Dictyostelium discoideum*. The amoebae grow as single cells, as long as food is present in the form of bacteria. When starved, the amoebae aggregate by a chemotactic response to signals of cyclic AMP (cAMP) emitted in a pulsatile manner by cells behaving as aggregation centers [22]. The signals are relayed toward the periphery of the aggregation territory by amoebae behaving as excitable cells. Like neurons, *Dictyostelium* amoebae behave either as autonomous oscillators or as excitable cells. The oscillations of cAMP were also observed in suspensions of *Dictyostelium* cells, with a period of the order of 5 min [23]. The oscillatory secretion of cAMP is associated with the formation of concentric or spiral waves of chemotactic movement when the cells are allowed to aggregate on agar [22]. This phenomenon still represents one of the most striking examples of spatiotemporal organization at the supracellular level [24, 25].

What is the function pulsatile of pulsatile cAMP signals in *Dictyostelium*? In the slime mold *Dictyostelium minutum*, aggregation occurs without oscillations or waves, because aggregation centers release the chemotactic factor in a monotonous manner. As a result, aggregation territories are much smaller in this species of slime mold. Thus, the existence of relay and oscillations of cAMP signals in *Dictyostelium discoideum* amoebae allows aggregation centers to control the aggregation of a much larger number of cells over a wider territory. The experimental studies brought to light an additional function of pulsatile release of cAMP signals, namely, their frequency encoding. Indeed, much as for GnRH in mammals, pulses of cAMP delivered too frequently or at too large intervals, or in a random manner fail to promote aggregation [26]. Modeling the periodic release of cAMP in *D. discoideum* shows that the pulsatile secretion provides the amoebae with a means to avoid the loss of target cell responsiveness due to receptor desensitization [27, 28]. A similar explanation holds for pulsatile hormonal signaling [29]. The optimal frequency is of one pulse every 5 min or every hour for cAMP and GnRH signaling, respectively. This frequency is governed by the time needed for resensitization of the cAMP or GnRH receptor after stimulation by a cAMP or hormone pulse.

A last example of the role of frequency coding, this time related to intracellular signaling, pertains to the role of Ca^{++} oscillations in development. At fertilization, the fusion of the spermatozoon with the oocyte triggers a train of Ca^{++} pulses. This occurs through the injection into the egg of a protein initially called "oscillin". Subsequent studies showed that this protein is in fact the isoform ζ of an enzyme, phospholipase C (PLC), that raises the level of inositol triphosophtate (IP_3), which in turn elicits the release of Ca^{++} from intracellular stores that eventually leads to the onset of Ca^{++} oscillations [30, 31]. The latter are required for the resumption of mitotic cycles and the associated development of the egg into an adult organism. Remarkably, certain cases of male infertility can be traced back to mutations of PLCζ that prevent the rise in IP_3 and the triggering of a sufficient number of repetitive Ca^{++} pulses needed for egg development [32].

Yet another function of biological rhythms is to allow for the formation of ordered spatial structures during development. A most striking example is provided by the segmentation clock, which underlies the periodic expression of genes involved in the formation of somites in vertebrate embryos [33]. A new pair of somites forms after each cycle of gene expression controlled by the segmentation clock, the period of which ranges from 30 min to several hours in different organisms [34]. Conspicuous of this phenomenon is the fact that a temporal oscillatory process gives rise to a spatial pattern during development [33].

While most biological rhythms possess a clear physiological role, the function of some of them remains unclear. A case in point is that of glycolytic oscillations, which remain to this day the prototype of oscillatory behavior in a metabolic pathway [7]. These oscillations were first observed some sixty years ago in yeast cell suspensions receiving a constant input of glucose [35]. The conversion of this glycolytic substrate into ethanol and CO_2 was found to oscillate with a period of several minutes, even though the substrate input remained fixed at a constant value. Oscillations occurred in a range of substrate input rates bounded by one (in intact yeast cells) or two (in yeast extracts) critical values [36]. The mechanism of the oscillations relies on the activation of phosphofructokinase (PFK), a key glycolytic enzyme, by a reaction product, ADP, or AMP [37].

The physiological significance of glycolytic oscillations in yeast remains an open question. Several hypotheses were presented over the years, which generally suggest an increased metabolic efficiency of this ATP-producing biochemical pathway. It may well be, however, that glycolytic oscillations do not perform any particular function in yeast cells and simply occur accidentally as a result of the specific regulation of PFK: at a local level, this enzyme utilizes ATP as substrate, while it is regulated according to the global role of glycolysis, which is to produce ATP. As a result, PFK is activated by its product ADP, and by AMP, which is linked to ADP by the enzyme adenylate kinase. The subsequent autocatalytic regulation is responsible for the instability that leads to glycolytic oscillations. The oscillations have also been observed in other cell types, e.g., pancreatic β cells, where they appear to be involved in insulin secretion [38].

3 Mechanisms of Biological Rhythms

While the variety of biological rhythms corresponds to widely different underlying mechanisms, there is a common principle that links them all, namely, regulatory feedback. Among the most important types of regulation involved in the mechanism of biological rhythms, we can distinguish those that occur within cells as a result of regulation of (i) gene expression by inducers or repressors, (ii) enzyme or receptor activity, or (iii) ion transport across membranes. Thus, the regulation of the intracellular concentration of a protein can be exerted through modulating the synthesis or/and degradation of its mRNA or of the protein itself. Regulation can take the form of positive or negative feedback loops, which often coexist in complex regulatory networks. Such multiple modes of regulation underlie a variety of cellular rhythms ranging from neural rhythms to metabolic oscillations, and from cell cycle oscillations to the circadian clock. At the supracellular level, regulations between cells are at work in cell differentiation, developmental processes, the immune response, or in neural networks. Regulations between different organs occur in neuro-endocrine signaling. Finally, at the level of interactions between organisms, regulations occur between different microbial or animal species, as exemplified by predator-prey interactions. The latter example was the subject of one of the earliest attempts to model biological oscillations mathematically [39].

Examples of feedback regulations leading to instabilities and oscillations include the autocatalytic regulation of the allosteric enzyme phosphofructokinase responsible for the onset of glycolytic oscillations in yeast and muscle cells [7, 40, 41], and oscillations of cAMP in *Dictyostelium* amoebae. In the latter case, extracellular cAMP binds to a cell surface receptor and thereby activates cAMP synthesis catalyzed by the membrane-bound enzyme adenylate cyclase; intracellular cAMP is transported into the extracellular medium, thus creating a positive feedback loop. This self-amplified process is limited by negative feedback in the form of desensitization of the cAMP receptor [28]. Positive feedback is also involved in intracellular Ca^{++} oscillations, which originate from the process of Ca^{++} -induced Ca^{++} release from intracellular stores [42, 43]. Here again, a limiting process, such as depletion of the intracellular stores, causes the decreasing phase of Ca^{++} pulses, much as substrate depletion causes the decreasing phase of glycolytic oscillations.

Circadian clocks represent a major example of biological rhythm, which allow the adaptation of all eukaryotic organisms and some bacterial species to the periodic nature of terrestrial environment [44, 45]. The oscillations provide another example of biological rhythm based on negative feedback. Experimental studies in the fly *Drosophila* showed that circadian oscillations of a period close to 24 h result from the negative feedback exerted by the clock proteins PER and TIM on the transcription of their genes [46]. The nature of the clock proteins involved in this negative feedback regulation may vary in different organisms such as *Neurospora* or mammals, but the principle of auto-regulatory negative feedback on gene transcription holds for the mechanism of circadian clocks in a wide variety of species [47]. While in mammals the circadian pacemaker resides in synchronized neurons of the suprachiasmatic nuclei in the hypothalamus [48], peripheral clocks exist in various organs, which oscillate at a circadian period with specific phases [49, 50]. Mathematical models for circadian rhythms based on negative autoregulation of gene expression were proposed first for *Drosophila* [51] and later for mammals [52–55].

The early cell cycles in amphibian embryos are driven by a biochemical oscillator involving the protein kinase cdc2 and a protein named cyclin [14]. The synthesis of cyclin leads to the formation of an active cyclin-cdc2 complex, which triggers mitosis and subsequently activates the enzymatic degradation of cyclin [56]. This negative feedback loop is at the core of the embryonic mitotic oscillator [57]. In mammalian cells, the periodic progression along the cell cycle phases G1, S (DNA replication), G2 and M (mitosis) is driven by a network of cyclin-dependent kinases (Cdks). Each Cdk module consisting of a particular Cdk and its associated cyclin, controls a specific phase of the cell cycle. Activation of a given Cdk module leads to the activation of the following module in the network and to the inactivation of the previous modules [15]. Multiple layers of regulation control the Cdk network at the level of cyclin synthesis and degradation and through Cdk activation or inhibition through phosphorylation-dephosphorylation. The dynamics of the Cdk network can be modeled as that of a self-sustained oscillator [58, 59], even if any particular Cdk module can give rise to bistable transitions [60–63] while the whole network of tightly coupled Cdk modules undergoes global oscillations [58, 64]. Multiple layers of regulation of the Cdk network lead to the transient, sequential activation of the Cdk modules and thereby ensure that the successive cell cycle phases occur in a repetitive, ordered manner.

Among biological rhythms, those that are based on the control of voltage gated ion channels are widespread and play key physiological roles [4, 13]. These rhythms underlie neural and cardiac oscillations, and a variety of brain rhythms involved, for example, in respiration and the sleep-wake cycle. Beyond occasional differences in the nature of the ions involved, these rhythms of electrical nature are based on the coupling of three successive phases characterizing the changes in the membrane potential of excitable cells: (1) an inward depolarizing current, often carried by Na^+ ions, is responsible for the depolarizing phase of the action potential; (2) a second phase corresponding to repolarization is due to an outward current, often carried by K^+ ions; (3) autonomous oscillations arise when a "pacemaker" current activated by hyperpolarization brings back the membrane potential above the excitation threshold, thus leading to the spontaneous production of a new action potential. Different excitable cells produce different types of action potential, with different periods and waveforms—for example, cardiac action potentials are characterized by a prolonged plateau of depolarization, due to the influx of Ca^{++} ions [12, 13].

Some excitable cells, exemplified by the neuron R15 in *Aplysia* [65], produce complex periodic oscillations of the "bursting" type in which active phases of high-frequency oscillations are separated by silent phases during which the action potential remains at a low, repolarized value. The mechanism underlying such complex periodic oscillations involves the interplay of a larger number of ionic conductances [66]. While simple or complex oscillations of the membrane potential can occur in isolated neurons or cardiac cells, they can also originate in neural networks from the interplay between different cells coupled through negative or positive interactions [67]. Oscillatory neural networks, referred to as "central pattern generators", are involved in many important physiological functions such as the control of locomotion or of gastro-intestinal activity [68, 69].

Beyond the differences in molecular mechanism and period, biological rhythms possess some common properties. The most important is that most of them are robust and correspond to the evolution to a limit cycle in phase space: regardless of initial conditions, the oscillatory system evolves in time to a unique closed curve characterized by a unique period and amplitude [7]. In some cases, more than one stable limit cycle can be reached, starting from a set of initial conditions that represent the basin of attraction of this particular limit cycle. Such cases of multi-rhythmicity have been observed in a number of theoretical models [70] and, so far, in but a few experimental studies, e.g. in the forcing of cardiac cells by periodic stimuli [71].

Feedback interactions are necessary for the occurrence of instabilities, but a second requirement pertains to the nonlinear nature of the positive or negative regulatory interactions. Such nonlinearity is readily achieved in biochemical processes, thanks to the cooperative nature of the conformational transitions of allosteric proteins observed for enzymes, receptors, transcription factors, repressors, transport proteins, or ion channels. Sigmoidal dose-response curves reflecting the existence of thresholds in cellular regulation can be generated not only through cooperative interactions between subunits of allosteric proteins [72] but also through the phenomenon of ultrasensitivity generated in biochemical cascades controlled by reversible protein phosphorylation or other modes of protein covalent modification [73].

While nonlinear feedback processes play a prominent role in the onset of instabilities leading to oscillations, theoretical studies showed that time delays in feedback loops favor the occurrence of instabilities and oscillatory behavior. In his modeling work on cell differentiation in hematopoiesis, Michael Mackey [74] has brought to light the role of time delays in the onset of oscillatory dynamics associated with a variety of hematological disorders [75].

Experimental studies have uncovered over the years the molecular or cellular mechanism for many if not most biological rhythms. Mathematical models were developed for these rhythms on the basis of these experimental findings. For some important physiological rhythms, the search for the underlying mechanism still goes on. A case in point is the segmentation clock that controls the periodic formation of somites during the embryonic development of vertebrate organisms [33]. The segmentation clock triggers the oscillatory expression of specific genes, with a period ranging from 30 min to several hours in different organisms. The phenomenon involves oscillations in the Notch, FGF and Wnt signaling pathways, which appear to be coupled [76]. Feedback loops within each of these signaling pathways are in principle capable of producing oscillations [77, 78], as observed experimentally, but the question remains as to whether some master biochemical oscillator drives the oscillations observed in the three signaling pathways.

In the last two decades, the study of biological rhythms has been extended to synthetic biology. Oscillatory gene regulatory circuits have been constructed in the laboratory, the prototype being the *Repressilator* in which three repressors are coupled in a cyclical manner [79]. Subsequent studies aimed at constructing oscillatory gene circuits whose period can be tuned at a particular value [80, 81]. Here also the experimental realization is guided by the theoretical modeling of a gene regulatory network based on negative feedback on gene transcription.

4   Disorders of Biological Rhythms: Dynamical Diseases

It is no wonder that in view of their key physiological functions, disturbances of biological rhythms often lead to physiological disorders. A major contribution of Michael Mackey and Leon Glass at the Physiology Department of McGill University was to stress this aspect and bring these disorders within the common framework of "dynamical diseases" [82, 83]. A decade later, in September 1986, in Bremen, together with Ludger Rensing and Uwe an der Heiden, Michael Mackey organized on the topic "Temporal Disorder in Human Oscillatory Systems" a seminal conference, which I was fortunate to attend. This conference showed [84] that in studying the mechanism and function of biological rhythms, it is crucial to investigate the nonlinear dynamics of associated physiological disorders.

One case in point is the heart rhythm for which a vast repertoire of dynamical disorders has been characterized, ranging from various patterns of arrhythmias to atrial and ventricular fibrillation. The response of oscillating cardiac cells to pulsatile stimulation was studied in detail at Mc Gill by Leon Glass and his colleagues, Michael Guevara and Alvin Shrier [71]. They analyzed mathematical models for this situation and carried out a bifurcation analysis, the results of which

Table 2 Some physiological disorders associated with biological rhythms [9]

Cardiac arrhythmias and fibrillation
Large-amplitude neuronal oscillations in epileptic seizures
Breathing pattern disorders (e.g., Cheyne-Stokes respiration)
Sleep disorders linked to the circadian clock (familial advanced or delayed sleep phase syndromes, jet lag, 12 h-phase shift of melatonin circadian rhythm in Smith-Magenis syndrome)
Scoliosis originating from alterations of the segmentation clock controlling somitogenesis
Morphogenetic defects associated with abnormal Ca^{++} oscillations (Noonan syndrome)
Deafness associated with impaired Ca^{++} wave propagation in cochlear cells
Premature puberty (premature onset of pulsatile GnRH secretion)
Female infertility due to lack of ovulation resulting from abnormal GnRH pulsatile secretion
Male infertility due to lack of sperm-induced Ca^{++} oscillations in fertilized egg
Premature delivery (premature onset of uterine contractions triggered by Ca^{++} oscillations)
Intestinal disorders associated with alteration or lack of periodic contractions
Muscular tremor (e.g., in Parkinson's disease)
Periodic hematological diseases (chronic myelogenous leukemia, cyclic thrombocytopenia, cyclic neutropenia)
Bipolar disorders (spontaneous alternation of phases of mania and depression)
Human weight cycling ("yo-yo" dieting)

could be related to the observed physiological disorders of cardiac function. Beyond the case of the heart, disorders are observed for a large variety of biological rhythms, as indicated in Table 2, which presents a non-exhaustive list of such physiological disorders; see [9] for further details and references to the original publications.

Many of these disturbances represent dynamical diseases, i.e., pathological modes of dynamical behavior that arise in cellular or supracellular control systems operating in non-physiological conditions [82, 83]. Chaotic behavior in a physiological system that normally oscillates periodically provides an example of such a situation. Conversely, dynamical diseases can take the form of periodic or chaotic oscillations occurring in a system that normally operates in a stable steady state.

Because biological rhythms generally occur in a well-defined domain in parameter space, it is not surprising that non-physiological behavior occurs as soon as the system quits the domain of oscillatory behavior following a change in some control parameter value. Many specific examples of such transitions were studied by means of mathematical modeling. Besides the above-mentioned case of cardiac oscillations subjected to pulsatile electrical stimulation, the work of Michael Mackey [74] on hematopoietic diseases exemplifies this approach. Many hematological disorders involving different types of blood cells indeed possess a cyclic nature, such as chronic myelogenous leukemia, cyclic thrombocytopenia, and cyclic neutropenia. An important contribution of Mackey and coworkers was to show that time delays in the maturation process of blood cells favor the onset of instabilities leading to oscillations [74, 75].

Another important example of physiological disorders associated with the dysfunction of biological rhythms pertains to disorders of pulsatile hormone secretion. In Sect. 2 we discussed the function of pulsatile secretion observed for many hormones, the prototype of which remains the secretion of GnRH by the hypothalamus with a frequency of one 5-min pulse every hour in the ewe and in the human female. An abnormal frequency of this secretion causes the failure to induce the secretion by the pituitary of the target hormones LH and FSH at levels appropriate for ovulation [19–21]. The frequency encoding of pulsatile hormone secretion is also observed for other hormones, e.g., the growth hormone, GH [85]. Regarding GnRH, it is important to note that other rhythms play key roles in the reproductive process. For example, Ca^{++} oscillations triggered at fertilization following the insertion of the enzyme PLCζ by the spermatozoon into the egg are needed for development of the egg [30, 31]. The lack of Ca^{++} oscillations due to a mutation in the enzyme PLCζ is a cause of infertility in men [32].

Disorders of the sleep-wake cycle represent another major class of physiological disturbance of a key biological rhythm. Some of these disorders can be related to the circadian clock [86]. The sleep-wake cycle is indeed driven by the circadian clock, which is governed by a complex regulatory network based on transcription-translation feedback loops involving a dozen proteins in mammals, such as PER, CRY, CLOCK, BMAL1 and REV-ERBα (see Sect. 3). As a result of mutations or alterations in cellular conditions, changes in the levels of these proteins may occur, due to increased or decreased rates of gene transcription, protein synthesis or protein degradation. Such changes may abolish endogenous circadian oscillations or may simply alter their amplitude or their phase with respect to the light-dark (LD) cycle to which it is entrained.

How the sleep-wake cycle is linked to the circadian clock remains to be fully clarified at the molecular level. Because the circadian clock is itself periodically driven by the light-dark cycle (owing to the enhanced rate of gene expression of the PER protein in the light phase), a change in the phase of circadian oscillations can produce a change in the phase of the sleep-wake cycle. This phase shift takes the form of either a phase advance or a phase delay. Both types of situations have been characterized clinically and are associated with mutations in the PER protein or in casein kinase 1ε, the enzyme by which PER is phosphorylated: the former situation corresponds to a syndrome known as the familial phase advanced sleep phase syndrome (FASPS), and the latter to the mirror familial phase delayed sleep phase syndrome (FDSPS) [87, 88]. The molecular basis of the two types of disturbance is, respectively, an enhanced or decreased phosphorylation of the PER protein that plays a primary role in the negative feedback loop at the core of the circadian clock (PER makes a complex with CRY and this complex inactivates the complex CLOCK-BMAL1 that induces the expression of the genes coding for PER and CRY). The change in the phase of the circadian clock is due to a change in its autonomous period: a decrease in PER phosphorylation leads to a decrease in period and to a phase advance of the clock, while a rise in PER phosphorylation causes the opposite effects [53].

Rather than altering the phase of the oscillations, dynamical disorders of the circadian clock may also affect its entrainment by the light-dark cycle. Thus, the study of a detailed molecular model for the mammalian circadian clock unveiled the possibility that in a certain parameter domain corresponding to a reduced level of CRY protein, the circadian clock mechanism fails to be entrained by the LD cycle. A sufficient increase in CRY level, due to enhanced synthesis or decreased degradation of the protein or its mRNA, readily restores the entrainment of the clock by the LD cycle [89]. Interestingly, the absence of entrainment of the circadian clock by the LD cycle corresponds for non-blind people to a syndrome clinically known as the non-24 h sleep-wake cycle syndrome. For people affected by this syndrome, the phase of the sleep-wake cycle is progressively drifting through the day or through the night, so that they go to sleep at variable times. The study of a theoretical model thus allows us to uncover one possible molecular mechanism leading to this disorder of the sleep-wake cycle, namely, the insufficient level of the CRY protein in the circadian clock network [89, 90].

Some dynamical disorders in humans pertain to rhythms that possess a psychological component. One example is that of bipolar disorders, which involve the cycling of mood between two states, mania and depression. The molecular and cellular mechanisms of this disorder remain unclear, but it is possible to approach the dynamics of this disorder by means of a theoretical framework based on two neural circuits regulated by mutual inhibition [91]. The two putative circuits promote mania and depression respectively. Such a regulatory pattern, common in neurobiology, has been invoked for the control of REM sleep [92]. The two-variable model for the dynamics of bipolar disorders [91] allows us to see how such a regulatory circuit can produce bistable transitions between mania and depression. The insertion of two additional variables in the network further gives rise to sustained oscillations between the two mood states.

Yet another rhythm is that of human weight cycling [93]. Modeling this phenomenon by means of a three-variable model shows that the psychological mechanism at work here also gives rise to sustained oscillations of the limit cycle type in a closed domain in parameter space [94]. Here again two distinct modes of dynamic behavior are observed in adjacent domains in parameter space: one corresponds to a stable steady state while the other is associated with sustained oscillations around an unstable steady state. For weight cycling as for bipolar disorders, oscillations could be viewed as a physiological disorder or a dynamical disease, while a stable healthy steady state would represent a physiologically desirable situation.

5   Why So Many Biological Rhythms?

After presenting an overview of biological rhythms, the physiological functions they perform, and the disorders associated with their malfunctions, let us return to the question raised at the outset: why is life associated with oscillations at all levels of biological organization?

Leaving aside oscillations in physical systems, oscillations have been observed outside a biological context in a few chemical reactions, the prototype of which remains the Belousov-Zhabotinsky reaction. But what distinguishes purely chemical oscillations from oscillations associated with biological rhythms? The main difference is that in contrast to purely chemical systems, biological systems are subjected to the constraint of selection that drives biological evolution. Regulation through feedback processes lies behind the selection pressure, at all levels of biological organization, from molecular interactions within cells to regulatory interactions between cells, organs, or animal populations. While self-inhibition or self-activation may occur accidentally, chemical systems do not involve any systematic regulation by feedback: this is the key feature that distinguishes them from biological systems in which regulatory feedback is needed to optimize the many physiological functions they perform. In addition to this role, regulatory feedback provides a major source of instability capable of producing oscillatory behavior. An additional factor that enhances the propensity of regulatory feedback to produce oscillations is the nonlinearity of biological processes at the molecular, cellular and supracellular levels. The sources of nonlinearity in biological systems are multifarious: they include regulation in the form of positive or negative feedback; cooperative processes involving allosteric proteins such as regulatory enzymes, receptors, or transcription factors; saturation of enzymes or receptors; and ultrasensitivity of biochemical systems controlled by reversible covalent modification, e.g., phosphorylation-dephosphorylation. Voltage gated ion transport across excitable membranes, which plays a key role in neural and cardiac oscillations, provides an additional source of nonlinearity, together with the activation or inhibition of ion conductances by neurotransmitters.

In each case, oscillations are the signature of an instability. While biochemical or biological systems often evolve, for a given set of environmental conditions, toward a stable steady state, feedback loops—as well as other contributing factors, such as time delays—may destabilize the steady state once the regulatory interactions are sufficiently nonlinear. The instability of the steady state generally occurs in a domain in parameter space bounded by critical values corresponding to bifurcation points beyond which sustained oscillations occur. In the phase plane, these oscillations correspond to a limit cycle surrounding the unstable steady state. While the above discussion primarily pertains to biological rhythms, it is important to stress that the same mechanisms that give rise to oscillations can often produce, in slightly different conditions, a variety of other nonlinear dynamical phenomena such as complex oscillations (including bursting or chaos), the coexistence between multiple steady states (bistability or tristability), or the coexistence between multiple rhythms (birhythmicity or trirhythmicity), as well as spatial patterns or waves [95, 96].

The multiplicity of feedback regulations and the nonlinear nature of biochemical and cellular processes explain why instabilities leading to oscillations are so widespread in biological systems. As soon as regulations provide a selective advantage to an organism, they are bound to be selected in the course of evolution. While some of these regulations allowed a better utilization of available nutrients and thereby improved metabolic efficiency, other regulations gave rise to oscillations. If the latter played a new physiological role which proved to be beneficial to the cellular or multicellular organism –e.g. neural or cardiac rhythms, circadian rhythms, cell cycle clocks, hormonal rhythms, the sleep-wake cycle, to name but a few of the rhythms listed in Table 1—such rhythm-inducing regulations were retained and further modulated by the selection pressure. For those rhythms whose function remains unclear, such as glycolytic oscillations in yeast, a function may still be found but the possibility remains that they emerged accidentally as a by-product of cellular regulation and were conserved if they were not detrimental to the cell.

In studying the mechanism and function of biological rhythms, mathematical models have long proved to be a most useful complement of experimental studies. The tools used in theoretical studies range from analytical results, in relatively simple systems, to numerical simulations. The time evolution of models for biological rhythms is often described by ordinary differential equations, or time-delay equations. The models can contain a limited number of variables, typically between 2 and 10, or a larger number of variables, up to dozens or more. All these models are complementary and possess specific virtues. The smaller the number of variables, the more amenable are the models to mathematical or numerical analysis. Once the number of variables and parameters becomes very large, it may become difficult to clarify precisely the mechanism of the rhythm, but the detailed theoretical description of the mechanism is more faithful and thereby allows us to thoroughly investigate the effect of a particular variable or regulatory interaction.

References

1. Burger, M., Bujdoso, E.: Oscillating chemical reactions as an example of the development of a subfield of science. In: Field, R.J., Burger, M. (eds.) Oscillations and Traveling Waves in Chemical Systems, pp. 565–604. Wiley-Blackwell, New York (1985)
2. Noyes, R.M., Field, R.J., Köros, E.: Oscillations in chemical systems. 1. Detailed mechanism in a system showing temporal oscillations. J. Am. Chem. Soc. **94**, 1394–1395 (1972)
3. Zhabotinsky, A.M.: Periodic process of the oxidation of malonic acid in solution. Study of the kinetics of Belousov's reaction. Biofizika. **9**, 1306 (1964)
4. Buzsaki, G.: Rhythms of the Brain. Oxford Univ. Press, New York (2006)
5. Fessard, A.: Propriétés rythmiques de la matière vivante. Hermann, Paris (1936)
6. Glass, L., Mackey, M.C.: From Clocks to Chaos: The Rhythms of Life. Princeton Univ. Press, Princeton (1988)
7. Goldbeter, A.: Biochemical Oscillations and Cellular Rhythms. The Molecular Bases of Periodic and Chaotic Behaviour. Cambridge University Press, Cambridge (1996)
8. Goldbeter, A.: La Vie oscillatoire. Odile Jacob, Paris (2010) (Revised and augmented edition: Au cœur des rythmes du vivant. La Vie oscillatoire. Paris: Odile Jacob; 2018)
9. Goldbeter, A.: Dissipative structures and biological rhythms. Chaos. **27**, 104612 (2017). https://doi.org/10.1063/1.4990783
10. Winfree, A.T.: The Geometry of Biological Time, 2nd edn. Springer, New York (2001)
11. Feldman, J., Del Negro, C.: Looking for inspiration: new perspectives on respiratory rhythm. Nat. Rev. Neurosci. **7**, 232–241 (2006). https://doi.org/10.1038/nrn1871
12. DiFrancesco, D.: Pacemaker mechanisms in cardiac tissue. Annu. Rev. Physiol. **55**, 455–472 (1993)
13. Noble, D.: The Initiation of the Heartbeat. Oxford Univ Press, Oxford (1979)

14. Murray, A., Hunt, T.: The Cell Cycle: An Introduction. W.H. Freeman and Company, New York (1993)
15. Morgan, D.O.: The Cell Cycle: Principles of Control. Oxford University Press, Oxford (2006)
16. Dehaene, S. : Le Code de la conscience. Odile Jacob, Paris (2014)
17. Leng, G. (ed.): Pulsatility in Neuroendocrine Systems. CRC Press, Boca Raton, FL (1988)
18. Karsch, F.J.: Central actions of ovarian steroids in the feedback regulation of pulsatile secretion of luteinizing hormone. Annu. Rev. Physiol. **49**, 365–382 (1987)
19. Knobil, E.: Patterns of hormone signals and hormone action. New Engl. J. Med. **305**, 1582–1583 (1987)
20. Pohl, C.R., Richardson, D.W., Hutchison, J.S., Germak, J.A., Knobil, E.: Hypophysiotropic signal frequency and the functioning of the pituitary-ovarian system in the rhesus monkey. Endocrinology. **112**, 2076–2080 (1983)
21. Leyendecker, G.L., Wildt, L., Hansmann, M.: Pregnancies following intermittent (pulsatile) administration of GnRH by means of a portable pump ("Zyklomat"): a new approach to the treatment of infertility in hypothalamic amenorrhea. J. Clin. Endocr. Metab. **51**, 1214–1216 (1980)
22. Alcantara, F., Monk, M.: Signal propagation during aggregation in the slime mould *Dictyostelium discoideum*. J. Gen. Microbiol. **85**, 321–334 (1974)
23. Gerisch, G., Wick, U.: Intracellular oscillations and release of cyclic AMP from *Dictyostelium* cells. Biochem. Biophys. Res. Commun. **65**, 364–370 (1975)
24. Lauzeral, J., Halloy, J. Goldbeter, A.: Desynchronization of cells on the developmental path triggers the formation of waves of cAMP during *Dictyostelium* aggregation. Proc. Natl. Acad. Sci. USA **94**, 9153–9158 (1997)
25. Goldbeter, A.: Oscillations and waves of cyclic AMP in *Dictyostelium*: a prototype for spatio-temporal organization and pulsatile intercellular communication. Bull. Math. Biol. **68**, 1095–1109 (2006)
26. Nanjundiah, V.: Periodic stimuli are more successful than randomly spaced ones for inducing development in *Dictyostelium discoideum*. Biosci. Rep. **8**, 571–577 (1988)
27. Li, Y.X., Goldbeter, A.: Frequency encoding of pulsatile signals of cyclic AMP based on receptor desensitization in *Dictyostelium* cells. J. Theor. Biol. **146**, 355–367 (1990)
28. Martiel, J.-L., Goldbeter, A.: A model based on receptor desensitization for cyclic AMP signaling in *Dictyostelium* cells. Biophys. J. **52**, 807–828 (1987)
29. Li, Y.X., Goldbeter, A.: Frequency specificity in intercellular communication. Influence of patterns of periodic signaling on target cell responsiveness. Biophys. J. **55**, 125–145 (1989)
30. Nomikos, M., Kashir, J., Swann, K., Lai, F.A.: Sperm PLCζ: from structure to Ca^{2+} oscillations, egg activation and therapeutic potential. FEBS Lett. **587**, 3609–3616 (2013)
31. Swann, K., Larman, M.G., Saunders, C.M., Lai, F.A.: The cytosolic sperm factor that triggers Ca^{2+} oscillations and egg activation in mammals is a novel phospholipase C: PLCzeta. Reproduction. **127**, 431–439 (2004)
32. Kashir, J., Konstantinidis, M., Jones, C., Lemmon, B., Lee, H.C., Hamer, R., Heindryckx, B., Deane, C.M., De Sutter, P., Fissore, R.A., Parrington, J., Wells, D., Coward, K.: A maternally inherited autosomal point mutation in human phospholipase C zeta (PLCζ) leads to male infertility. Hum. Reprod. **27**, 222–231 (2012)
33. Pourquié, O.: The segmentation clock: converting embryonic time into spatial pattern. Science. **301**, 328–330 (2003)
34. Aulehla, A., Pourquié, O.: Oscillating signaling pathways during embryonic development. Curr. Opin. Cell Biol. **20**, 632–637 (2008)
35. Chance, B., Schoener, B., Elsaesser, S.: Control of the waveform of oscillations of the reduced pyridine nucleotide level in a cell-free extract. Proc. Natl. Acad. Sci. USA. **52**, 337–341 (1964)
36. Hess, B., Boiteux, A.: Oscillatory phenomena in biochemistry. Annu. Rev. Biochem. **40**, 237–258 (1971)
37. Madsen, M.F., Danø, S., Sørensen, P.G.: On the mechanisms of glycolytic oscillations in yeast. FEBS J. **272**, 2648–2660 (2005)

38. Bertram, R., Satin, L., Zhang, M., Smolen, P., Sherman, A.: Calcium and glycolysis mediate multiple bursting modes in pancreatic islets. Biophys. J. **87**, 3074–3087 (2004)
39. Volterra, V.: Fluctuations in the abundance of a species considered mathematically. Nature. **118**, 558–560 (1926)
40. Boiteux, A., Goldbeter, A., Hess, B.: Control of oscillating glycolysis of yeast by stochastic, periodic and steady source of substrate: a model and experimental study. Proc. Natl. Acad. Sci. USA. **72**, 3829–3833 (1975)
41. Goldbeter, A., Lefever, R.: Dissipative structures for an allosteric model. Application to glycolytic oscillations. Biophys. J. **12**, 1302–1315 (1972)
42. Dupont, G., Falcke, M., Kirk, V., Sneyd, J.: Models of Calcium Signalling. Springer International Publishing, Cham (2016)
43. Goldbeter, A., Dupont, G., Berridge, M.J.: Minimal model for signal-induced Ca^{2+} oscillations and for their frequency encoding through protein phosphorylation. Proc. Natl. Acad. Sci. USA. **87**, 1461–1465 (1990)
44. Bünning, E.: The Physiological Clock. Springer, Berlin-Heidelberg (1964)
45. Moore-Ede, M.C., Sulzman, F.M., Fuller, C.A.: The Clocks that Time Us: Physiology of the Circadian Timing System. Harvard University Press, Cambridge, MA (1982)
46. Hardin, P.E., Hall, J.C., Rosbash, M.: Feedback of the *Drosophila* period gene product on circadian cycling of its messenger RNA levels. Nature. **34**, 536–540 (1990)
47. Ukai, H., Ueda, H.R.: Systems biology of mammalian circadian clocks. Annu. Rev. Physiol. **72**, 579–603 (2010)
48. Yamaguchi, S., Isejima, H., Matsuo, T., Okura, R., Yagita, K., Kobayashi, M., Okamura, H.: Synchronization of cellular clocks in the suprachiasmatic nucleus. Science. **302**, 1408–1412 (2003)
49. Dibner, C., Schibler, U., Albrecht, U.: The mammalian circadian timing system: organization and coordination of central and peripheral clocks. Annu. Rev. Physiol. **72**, 517–549 (2010)
50. Mohawk, J.A., Green, C.B., Takahashi, J.S.: Central and peripheral circadian clocks in mammals. Annu. Rev. Neurosci. **35**, 445–462 (2012)
51. Goldbeter, A.: A model for circadian oscillations in the *Drosophila period* protein (PER). Proc. R. Soc. Lond. B. **26**, 319–324 (1995)
52. Forger, D.B., Peskin, C.S.: A detailed predictive model of the mammalian circadian clock. Proc. Natl. Acad. Sci. USA. **100**, 14806–14811 (2003)
53. Leloup, J.-C., Goldbeter, A.: Toward a detailed computational model for the mammalian circadian clock. Proc. Natl. Acad. Sci. USA. **100**, 7051–7056 (2003)
54. Mirsky, H.P., Liu, A.C., Welsh, D.K., Kay, S.A., Doyle 3rd., F.J.: A model of the cell-autonomous mammalian circadian clock. Proc. Natl. Acad. Sci. USA. **106**, 11107–11112 (2009)
55. To, T.L., Henson, M.A., Herzog, E.D., Doyle 3rd, F.J.: A molecular model for intercellular synchronization in the mammalian circadian clock. Biophys. J. **92**, 3792–3803 (2007)
56. Félix, M.-A., Labbé, J.-C., Dorée, M., Hunt, T., Karsenti, E.: Triggering of cyclin degradation in interphase extracts of amphibian eggs by cdc2 kinase. Nature. **346**, 379–382 (1990)
57. Goldbeter, A.: A minimal cascade model for the mitotic oscillator involving cyclin and cdc2 kinase. Proc. Natl. Acad. Sci. USA. **88**, 9107–9111 (1991)
58. Gérard, C., Goldbeter, A.: Temporal self-organization of the cyclin/Cdk network driving the mammalian cell cycle. Proc. Natl. Acad. Sci. USA. **106**, 21643–21648 (2009)
59. Gérard, C., Goldbeter, A.: The balance between cell cycle arrest and cell proliferation: control by the extracellular matrix and by contact inhibition. Interface Focus. **4**, 20130075 (2014)
60. Mochida, S., Rata, S., Hino, H., Nagai, T., Novak, B.: Two bistable switches govern M phase entry. Curr. Biol. **26**, 3361–3367 (2016)
61. Novák, B., Tyson, J.J.: Numerical analysis of a comprehensive model of M-phase control in *Xenopus* oocyte extracts and intact embryos. J. Cell Sci. **106**, 1153–1168 (1993)
62. Sha, W., Moore, J., Chen, K., Lassaleta, A.D., Yi, C.-S., Tyson, J.J., Sible, J.C.: Hysteresis drives cell-cycle transitions in *Xenopus laevis* egg extracts. Proc. Natl. Acad. Sci. USA **100**, 975–980 (2003)

63. Pomerening, J.R., Sontag, E.D., Ferrell Jr., J.E.: Building a cell cycle oscillator: hysteresis and bistability in the activation of Cdc2. Nat. Cell Biol. **5**, 346–351 (2003)

64. Gérard, C., Gonze, D., Goldbeter, A.: Effect of positive feedback loops on the robustness of oscillations in the network of cyclin-dependent kinases driving the mammalian cell cycle. FEBS J. **279**, 3411–3431 (2012)

65. Alving, B.O.: Spontaneous activity in isolated somata of *Aplysia* pacemaker neurons. J. Gen. Physiol. **51**, 29–45 (1968)

66. Adams, W.B., Benson, J.A.: The generation and modulation of endogenous rhythmicity in the *Aplysia* bursting pacemaker neurone R15. Progr. Biophys. Mol. Biol. **46**, 1–49 (1985)

67. Selverston, A.I., Moulins, M.: Oscillatory neural networks. Annu. Rev. Physiol. **47**, 29–48 (1985)

68. Grillner, S., Wallén, P.: Central pattern generators for locomotion, with special reference to vertebrates. Annu. Rev. Neurosci. **8**, 233–261 (1985)

69. Marder, E., Bucher, D.: Central pattern generators and the control of rhythmic movements. Curr. Biol. **11**, 986–996 (2001)

70. Goldbeter, A., Yan, J.: Multi-synchronization and other patterns of multi-rhythmicity in oscillatory biological systems. Interface Focus. **12**, 20210089 (2022)

71. Guevara, M.R., Shrier, A., Glass, L.: Chaotic and complex cardiac rhythms. In: Zipes, D.P., Jalife, J. (eds.) Cardiac Electrophysiology: from Cell to Bedside, pp. 192–200. WB Saunders, Philadelphia, PA (1990)

72. Monod, J., Wyman, J., Changeux, J.-P.: On the nature of allosteric transitions: a plausible model. J. Mol. Biol. **12**, 88–118 (1965)

73. Goldbeter, A., Koshland Jr., D.E.: An amplified sensitivity arising from covalent modification in biological systems. Proc. Natl. Acad. Sci. USA. **78**, 6840–6844 (1981)

74. Mackey, M.C.: Unified hypothesis for the origin of aplastic anemia and periodic hematopoiesis. Blood. **51**, 941–956 (1978) PMID: 638253

75. Foley, C., Mackey, M.C.: Dynamic hematological disease: a review. J. Math. Biol. **58**, 285–322 (2009)

76. Dequéant, M.L., Glynn, E., Gaudenz, K., Wahl, M., Chen, J., Mushegian, A., Pourquié, O.: A complex oscillating network of signaling genes underlies the mouse segmentation clock. Science. **31**, 1595–1598 (2006)

77. Goldbeter, A., Pourquié, O.: Modeling the segmentation clock as a network of coupled oscillations in the Notch, Wnt and FGF signaling pathways. J. Theor. Biol. **252**, 574–585 (2008)

78. Rodríguez-González, J.G., Santillán, M., Fowler, A.C., Mackey, M.C.: The segmentation clock in mice: interaction between the Wnt and Notch signalling pathways. J. Theor. Biol. **248**, 37–47 (2007)

79. Elowitz, M.B., Leibler, S.: A synthetic oscillatory network of transcriptional regulators. Nature. **403**, 335–338 (2000)

80. Stricker, J., Cookson, S., Bennett, M.R., Mather, W.H., Tsimring, L.S., Hasty, J.: A fast, robust and tunable synthetic gene oscillator. Nature. **456**, 516–519 (2008)

81. Tigges, M., Marquez-Lago, T.T., Stelling, J., Fussenegger, M.: A tunable synthetic mammalian oscillator. Nature. **457**, 309–312 (2009)

82. Glass, L., Mackey, M.C.: Pathological conditions resulting from instabilities in physiological control systems. Ann. N. Y. Acad. Sci. **316**, 214–235 (1979)

83. Mackey, M.C., Glass, L.: Oscillations and chaos in physiological control systems. Science. **197**, 287–289 (1977)

84. Rensing, L., an der Heiden, U., Mackey, M.C. (eds.): Temporal Disorder in Human Oscillatory Systems. Springer, Berlin (1987)

85. Hindmarsh, P.C., Stanhope, R., Preece, M.A., Brook, C.G.D.: Frequency of administration of growth hormone – An important factor in determining growth response to exogenous growth hormone. Hormone Res. **33**(Suppl. 4), 83–89 (1990)

86. Richardson, G.S., Malin, H.V.: Circadian rhythm sleep disorders: pathophysiology and treatment. J. Clin. Neurophysiol. **13**, 17–31 (1996)

87. Patke, A., Murphy, P.J., Onat, O.E., Krieger, A.C., Özçelik, T., Campbell, S.S., Young, M.W.: Mutation of the human circadian clock gene CRY1 in familial delayed sleep phase disorder. Cell. **169**, 203–215.e13 (2017). https://doi.org/10.1016/j.cell.2017.03.027
88. Toh, K.L., Jones, C.R., He, Y., Eide, E.J., Hinz, W.A., Virshup, D.M., Ptacek, L.J., Fu, Y.H.: An hPer2 phosphorylation site mutation in familial advanced sleep phase syndrome. Science. **291**, 1040–1043 (2001)
89. Leloup, J.-C., Goldbeter, A.: Modeling the circadian clock: from molecular mechanism to physiological disorders. BioEssays. **30**, 590–600 (2008)
90. Goldbeter, A., Leloup, J.-C.: From circadian clock mechanism to sleep disorders and jet lag: insights from a computational approach. Biochem. Pharmacol. **191**, 114482 (2021)
91. Goldbeter, A.: A model for the dynamics of bipolar disorders. Progr. Biophys. Mol. Biol. **105**, 119–127 (2011)
92. Lu, J., Sherman, D., Devor, M., Saper, C.B.: A putative flip-flop switch for control of REM sleep. Nature. **441**, 589–594 (2006)
93. Brownell, K.D.: Weight cycling. Am. J. Clin. Nutrition. **49**, 937 (1989)
94. Goldbeter, A.: A model for the dynamics of human weight cycling. J. Biosci. **31**, 101–108 (2006)
95. Goldbeter, A.: Dissipative structures in biological systems: bistability, oscillations, spatial patterns and waves. Phil. Trans. R. Soc. A. **376**, 20170376 (2018)
96. Tyson, J.J., Albert, R., Goldbeter, A., Ruoff, P., Sible, J.: Biological switches and clocks. J. R. Soc. Interface. **5**(Suppl. 1), S1–S8 (2008)

Mathematical Modeling of Heterogeneous Stem Cell Regeneration: From Cell Division to Waddington's Epigenetic Landscape

Jinzhi Lei ⓘ

Abstract Stem cell regeneration is a crucial biological process for most self-renewing tissues during the development and maintenance of tissue homeostasis. In developing the mathematical models of stem cell regeneration and tissue development, cell division is the core process connecting different scale biological processes and leading to changes in cell population number and the epigenetic state of cells. This chapter focuses on the primary strategies for modeling cell division in biological systems. The Lagrange coordinate modeling approach considers gene network dynamics within each cell and random changes in cell states and model parameters during cell division. In contrast, the Euler coordinate modeling approach formulates the evolution of cell population numbers with the same epigenetic state via a differential-integral equation. These strategies focus on different scale dynamics, respectively, and result in two methods of modeling Waddington's epigenetic landscape: the Fokker-Planck equation and the differential-integral equation approaches. The differential-integral equation approach formulates the evolution of cell population density based on simple assumptions in cell proliferation, apoptosis, differentiation, and epigenetic state transitions during cell division. Moreover, machine learning methods can establish low-dimensional macroscopic measurements of a cell based on single-cell RNA sequencing data. The low dimensional measurements can quantify the epigenetic state of cells and become connections between static single-cell RNA sequencing data with dynamic equations for tissue development processes. The differential-integral equation presented in this chapter provides a reasonable approach to understanding the complex biological processes of tissue development and tumor progression.

J. Lei (✉)
School of Mathematical Sciences, Center for Applied Mathematics, Tiangong University, Tianjin, China
e-mail: jzlei@tiangong.edu.cn

© The Author(s), under exclusive license to Springer Nature Switzerland AG 2025   37
Y. Mori et al. (eds.), *Dynamics of Physiological Control*, Lecture Notes
on Mathematical Modelling in the Life Sciences,
https://doi.org/10.1007/978-3-031-82396-1_4

1   Introduction

Biological processes are essentially multiscale dynamics at molecular, cellular, and tissue levels [1]. Individual cells usually show high heterogeneity at the molecular level with different expressions of genes and molecular interactions. At the cellular level, different types of cells undergo self-renew, differentiation, cell death, and migration; these behaviors are well-regulated to achieve a dynamic equilibrium in a tissue microenvironment. Biologically, the crosstalk between different scale dynamics is essential to maintain tissue growth. The multiscale biological processes occur at different spatial ranges and time scales. Mathematically, it is challenging to integrate the multiscale processes into a unified mathematical formulation to form a reasonable understanding of the biological process [2].

Cell division is the core that connects the molecular and tissue-level development processes and maintains tissue homeostasis. In cellular regeneration, the genetic regulatory network underlying cell cycling regulates the decision between cell growth, differentiation, and cell division. The molecular level dynamics within a cell determine the signals that trigger the irreversible proliferation process. Moreover, cell divisions include molecular processes of DNA replication, epigenetic modifications reallocation/reconstruction, protein synthesis, and the partitioning of molecules at cell division. These processes are stochastic biochemical reactions that may result in variations in the newborn cells. At the macroscale level, cell divisions result in changes in both the cell population and the density of different phenotypes of cells. Therefore, it is crucial to formulate cell division in mathematical models of biological systems in which microscale variations in each cell are considered.

During the maintenance of tissue homeostasis, adult stem cells undergo cell divisions to replace dying cells and regenerate damaged tissues through controlled self-renewal and differentiation [3]. Understanding the mechanisms that govern cell fate decisions and the regulation of self-renewal and differentiation in stem cells is of utmost significance. Waddington's epigenetic landscape has been fundamental in understanding cell fate decisions and differentiation [4]. However, while the Waddington landscape provides an intuitive understanding of the biological process, the mechanisms driving cell fate decisions still need to be discovered [5–7]. Specifically, there are still debates on the mathematical formulations of Waddington landscapes associated with different types of biological processes [8–11].

This chapter reviews the mathematical models of cellular regeneration from micro to macro scales and introduces a general mathematical framework that integrates various biological processes involved at multiple scales. With the help of this framework, we propose a dynamic equation that describes the evolution of the Waddington landscape for the development of a multiple-cell system.

Here, we briefly summarize the key formulations and ideas presented in this review.

At the microscale dynamics in a cell, the concentrations of functional molecules can be represented by a vector $\mathbf{x}$. The dynamics of the concentration of function molecules, $\mathbf{x}(t)$, is often described by a deterministic chemical rate equation of the form

$$\frac{d\mathbf{x}}{dt} = \mathbf{F}(\mathbf{x}). \tag{1}$$

Alternatively, when noisy fluctuations are considered, we have a stochastic equation of the form

$$\frac{d\mathbf{x}}{dt} = \mathbf{F}(\mathbf{x}) + \eta(t), \tag{2}$$

where $\eta(t)$ represented the random fluctuation in the changes in molecule concentrations.

The function $\mathbf{F}$ describes the regulatory relationships among functional molecules, such as gene regulatory networks, protein-protein interactions, or modification of molecules. These regulations may affect the biochemical interactions involved in the molecules' production and degradation/dilution. The reaction rates are often represented by parameters $\mathbf{q}$ that are involved in the equation. Hence, the Eq. (1) can be rewritten as (see Sect. 2)

$$\frac{d\mathbf{x}}{dt} = \mathbf{F}(\mathbf{x}; \mathbf{q}) \tag{3}$$

with parameter $\mathbf{q}$ explicitly included. Biologically, the parameter values $\mathbf{q}$ are usually non-constants and are dynamically changing over time. Thus, the parameters $\mathbf{q}$ should be represented as $\mathbf{q}(t)$.

During cell division, epigenetic modifications (histone modifications, DNA methylations, etc.) are redistributed and re-established at the daughter cells, followed by the reallocation of proteins and mRNAs. During cell division, these processes result in discontinuous changes in the variables $\mathbf{x}$ and the parameters $\mathbf{q}$. Therefore, the above equation should be extended below over cell divisions

$$\begin{cases} \dfrac{d\mathbf{x}}{dt} = \mathbf{F}(\mathbf{x}; \mathbf{q}(t)), & \text{Between cell divisions} \\ (\mathbf{x}, \mathbf{q}) \mapsto (\mathbf{x}', \mathbf{q}') \sim \mathcal{P}(\mathbf{x}', \mathbf{q}' | \mathbf{x}, \mathbf{q}), & \text{Cell division} \end{cases} \tag{4}$$

Here, $\mathcal{P}(\cdot | \mathbf{x}, \mathbf{q})$ represents the random numbers with a conditional distribution that is dependent on $\mathbf{x}$ and $\mathbf{q}$. The equation of form (4) provides a general framework for describing the dynamics of a single cell at a time scale across cell divisions.

Cell behaviors such as cell division, apoptosis, and differentiation/aging must be considered to formulate the population dynamics of multiple cells across cell divisions. Moreover, we should also consider the heterogeneity of cells due to the variance in epigenetic states and the transition between epigenetic states during cell divisions. Let $Q(t, \mathbf{x})$ represent the number of cells with epigenetic state $\mathbf{x}$, and the evolution of $Q(t, \mathbf{x})$ can be modeled with the following differential-integral equation (see Sect. 4.2)

$$
\begin{cases}
\dfrac{\partial Q(t, \mathbf{x})}{\partial t} = -Q(t, \mathbf{x})(\beta(c(t), \mathbf{x}) + \kappa(\mathbf{x})) \\
\qquad\qquad + 2 \displaystyle\int_{\Omega} \beta(c(t - \tau(\mathbf{y})), \mathbf{y}) Q(t - \tau(\mathbf{y}), \mathbf{y}) e^{-\mu(\mathbf{y})\tau(\mathbf{y})} p(\mathbf{x}, \mathbf{y}) d\mathbf{y}, \\
c(t) = \displaystyle\int_{\Omega} Q(t, \mathbf{x}) \zeta(\mathbf{x}) d\mathbf{x}.
\end{cases}
\tag{5}
$$

Here β, κ, μ represent the rates of proliferation, differentiation/senescence, and apoptosis, respectively; τ represents the duration of the proliferation phase; ζ represents the rate of cytokine secretion; c stands for the concentration of growth factors secreted by all cells; $p(\mathbf{x}, \mathbf{y})$ quantifies the transition probability of epigenetic states during cell division (Eq. (29) in Sect. 4.2).

There are two methods to formulate the evolution of the Waddington landscape. Consider the microscale dynamics formulated by Eq. (2), where $\eta = (\eta_1, \eta_2, \cdots, \eta_n)$ is a multi-dimensional Gaussian noise term, and the correlation satisfies $\langle \eta_i(t_1, \mathbf{x}) \eta_j(t_2, \mathbf{x}) \rangle = 2D\delta_{i,j}\delta(t_1 - t_2)$, with D as the diffusion coefficient. Let $P(t, \mathbf{x})$ represent the probability density of a cell in a state $\mathbf{x}$. Then, $P(t, \mathbf{x})$ satisfies the Fokker-Planck equation

$$
\frac{\partial P(t, \mathbf{x})}{\partial t} = \nabla \cdot (D\nabla P - \mathbf{F}P).
\tag{6}
$$

Moreover, if we introduce a birth-death rate $R(\mathbf{x})$ of a cell with state $\mathbf{x}$ and replace the probability density with the population density $f(t, \mathbf{x})$, the population density $f(t, \mathbf{x})$ satisfies the following population balance equation

$$
\frac{\partial f}{\partial t} = \nabla \cdot (D\nabla f) - \nabla \cdot (f\mathbf{F}) + Rf.
\tag{7}
$$

The potential

$$
U(t, \mathbf{x}) = -\log f(t, \mathbf{x})
\tag{8}
$$

give a formulation of the Waddington landscape. Detailed discussions are given in Sect. 7.

Alternatively, through Eq. (5), the evolution of the total cell number is given by

$$
Q(t) = \int_{\Omega} Q(t, \mathbf{x}) d\mathbf{x},
$$

and the population density of cells with epigenetic state $\mathbf{x}$ is represented as

$$f(t, \mathbf{x}) = \frac{Q(t, \mathbf{x})}{Q(t)}.$$

From Eq. (5), the evolution equation of $f(t, \mathbf{x})$ is obtained as (Eq. (57) in Sect. 7.2)

$$
\begin{aligned}
\frac{\partial f(t, \mathbf{x})}{\partial t} = {} & \frac{2}{Q(t)} \int_\Omega \beta(c_{\tau(\mathbf{y})}, \mathbf{y}) Q(t - \tau(\mathbf{y}), \mathbf{y}) e^{-\mu(\mathbf{y})\tau(\mathbf{y})} (p(\mathbf{x}, \mathbf{y}) - f(t, \mathbf{x})) d\mathbf{y} \\
& - f(t, \mathbf{x}) \int_\Omega f(t, \mathbf{y}) \left((\beta(c, \mathbf{x}) + \kappa(\mathbf{x})) - (\beta(c, \mathbf{y}) + \kappa(\mathbf{y})) \right) d\mathbf{y}.
\end{aligned}
$$

$$(9)$$

Equation (9) describes the evolution dynamics of the population density, or the Waddington landscape according to Eq. (8), of a multicellular system when heterogeneity and plasticity are considered.

The right-hand side of Eq. (9) represents the growth of the population density $f(t, \mathbf{x})$. We denote it as the growth operator $\mathcal{R}[f]$. Referring to Eq. (7), we can express this growth in the form of the population balance equation (see Sect. 7.3)

$$\frac{\partial f}{\partial t} = \nabla \cdot (D \nabla f) - \nabla \cdot (f \mathbf{F}) + \mathcal{R}[f]. \tag{10}$$

Equation (10) provides a formulation for the evolution of the population density that integrates both molecular-level dynamics in a cell and cell population dynamics. The population dynamics $Q(t)$ and the population density $f(t, \mathbf{x})$ together provide macroscopic descriptions of the population dynamics of multiple cellular systems with cell heterogeneity and plasticity.

2 Modeling the Cell Cycle

The cell cycle is a highly regulated and orchestrated series of events that can be divided into four main phases: G1 (Gap 1), S (Synthesis), G2 (Gap 2), and M (Mitosis). During G1, the cell prepares for DNA synthesis; during the S phase, DNA replication occurs; and in G2, the cell prepares for mitosis. Finally, during the M phase, the cell undergoes mitosis, wherein the duplicated chromosomes are equally distributed between the two daughter cells. A complex network of regulatory proteins, checkpoints, and feedback loops tightly regulates the transition from one phase to another.

Hundreds of mathematical models for the cell cycle have been published [12]. These mathematical models describe the rate of change of different molecular species over time. Ordinary differential equations (ODEs) are the most common mathematical framework in cell cycling modeling. If $\mathbf{x}$ represents the concentrations of particular molecules involved in the cell-cycling regulation circuit, the ODE would be expressed as

$$\frac{d\mathbf{x}}{dt} = \mathbf{F}(\mathbf{x}), \tag{11}$$

the function $\mathbf{F}$ is determined by how the molecules interact with each other to form autonomous oscillators in cell cycling.

For instance, a simple protein circuit in the in *Xenopus* embryos' cell cycle centers on cyclin-dependent protein kinase (CDK1), the anaphase-promoting complex (APC), and a protein like Polo-like kinase 1 (Plk1). In this circuit, CDK1 activates APC through Plk1, and APC inactivates CDK1 to form a negative feedback. Let [CDK1*], [APC*], and [Plk1*] represent the concentrations of active CDK1, APC, and Plk1, respectively. The interactions described result in the following two ODEs model [12]:

$$\begin{cases} \dfrac{d[CDK1^*]}{dt} = a_1 - b_1[CDK1^*]\dfrac{[APC^*]^{n_1}}{K_1^{n_1} + [APC1^*]^{n_1}}, \\[2mm] \dfrac{d[Plk1^*]}{dt} = a_2(1 - [Plk1^*])\dfrac{[CDK1^*]^{n_2}}{K_2^{n_2} + [CDK1^*]^{n_2}} - b_2[Plk1^*], \\[2mm] \dfrac{d[APC^*]}{dt} = a_3(1 - [APC^*])\dfrac{[Plk1^*]^{n_3}}{K_3^{n_3} + [Plk1^*]^{n_3}} - b_3[APC^*]. \end{cases} \tag{12}$$

Here, the parameters a_i represent the maximum protein production/activation rates, while b_i represent the degradation/inactivation rates. The total concentrations of active and inactive APC and Plk1 are assumed to be constant and normalized to 1. Equation (12) provides an example of Eq. (11) with $\mathbf{x} = ([CDK1^*], [APC^*], [Plk1^*])$, and $\mathbf{F}$ is defined by the right-hand side of (12). Sustained oscillatory dynamics can be formed with a proper selection of model parameters.

In cell-cycle modeling, we often consider constant model parameters for simplicity. However, the model parameters change with time for different reasons, such as cell growth and extracellular perturbations. Thus, the regulation function $\mathbf{F}$ often depends on time-dependent parameters $\mathbf{q}(t)$, and the equation of form (11) can be expressed as

$$\frac{d\mathbf{x}}{dt} = \mathbf{F}(\mathbf{x}; \mathbf{q}). \tag{13}$$

For instance, the gene expression rates $a_i, (i = 1, 2, 3)$ may be subjected to fluctuations in the intracellular microenvironment and epigenetic modification, and hence $\mathbf{q} = (a_1, a_2, a_3)$. Given the initial condition $\mathbf{x}(t_0) = \mathbf{x}_0$ and the temporal dynamics $\mathbf{q}(t)$, we can solve the Eq. (13) to obtain the time course of the system state $\mathbf{x}(t)$ for $t > t_0$.

The above time course is valid only within one cell cycle, and the dynamics across cell division must be considered to describe the long-term dynamics. Discontinuous changes in molecule concentration may happen following mitosis, by which molecules in a cell are redistributed to the two daughter cells. Moreover, epigenetic markers, including histone modifications and DNA methylations, are re-established for the newborn cells following DNA replication during the S phase of cell cycling. Thus, following cell divisions, a mother cell divides into two daughter cells, both the system state $\mathbf{x}$ and the parameter $\mathbf{q}$ may undergo discontinuous changes at the end of mitosis. We assume that the state $\mathbf{x}$ and the parameter $\mathbf{q}$ change following biological regulations with random perturbations. Hence, their values after mitosis can be considered as random numbers following a conditional distribution that depends on their mother cell values. The conditional distribution is denoted as $\mathcal{P}(\cdot|\mathbf{x}, \mathbf{q})$. Thus, we can extend Eq. (13) to a discontinuous dynamical equation across cell division as

$$
\begin{cases}
\dfrac{d\mathbf{x}}{dt} = \mathbf{F}(\mathbf{x}; \mathbf{q}), & \text{Between cell divisions} \\
(\mathbf{x}, \mathbf{q}) \mapsto (\mathbf{x}', \mathbf{q}') \sim \mathcal{P}(\mathbf{x}', \mathbf{q}'|\mathbf{x}, \mathbf{q}), & \text{Cell division}
\end{cases}
\tag{14}
$$

Biologically, random changes in the state $\mathbf{x}$ are often associated with the perturbation in the partition of molecules at cell division [13]. Let V_d and V_b be the cell volume at division and newborn, respectively. For the component x_i, the total number of $N_i = V_d x_i$ molecules are partitioned into two daughter cells, and the molecule number at one newborn cell is $V_b x_i'$. Let $r_i = V_b x_i'/V_d x_i$ represent the fraction of molecules allocated to one daughter cell, then $0 < r_i \leq 1$. Mathematically, the distribution of molecules partitioned into two daughter cells can be characterized by a binomial distribution random number, with the partition rate following a conjugate prior distribution–the beta distribution. Consequently, we can assume that r_i follows a beta distribution, providing a rule for how x_i changes during cell division.

Epigenetic regulations, such as histone modifications or DNA methylations, can interfere with chromatin structure and alter the gene expression rates. Hence, the expression rates a_i depend on each gene's epigenetic modification state u_i. The epigenetic state of each gene can refer to the fractions of marked nucleosomes or methylated CpG sites in the DNA segment of interest. Since the epigenetic states primarily affect the chromatin structure, they might influence the chemical potential required to initiate the transcription process. Thus, the expression rates a_i in (12) can be expressed as

$$
a_i = \alpha_i e^{\lambda_i u_i}, \quad i = 1, 2, 3.
$$

Here, λ_i represents the impact of the epigenetic modifications on the expression rates. Specifically, $\lambda_i > 0$ indicates an epigenetic modification that enhances the strength of cell activation, while $\lambda_i < 0$ indicates a modification that reduces this strength. The epigenetic state u_i may experience random changes during cell divisions, leading to corresponding random fluctuations in the expression rates a_i. For example, when u_i denotes the fraction of marked nucleosomes or methylated CpG sites, it lies within the $0 < u_i < 1$ range. Following a similar rationale as discussed for the distribution of r_i, we can posit that the value of u_i follows a beta distribution, with shape parameters depending on the cell state before cell division.

Given the rule of how **x** and **q** may change at mitosis, solving Eq. (14) gives the time course $(\mathbf{x}(t), \mathbf{q}(t))$ across multiple cell cycles, which corresponds to a long-term tracking of a cell over numerous cell cycles simulation. Nevertheless, (14) only describes the microscale dynamics inside a cell and cannot model the population dynamics of multiple cells.

3 Modeling Homogeneous Stem Cell Regeneration

It is crucial to understand how a system consisting of multiple cells evolves. To study the dynamics of stem cell regeneration, we employ the G0 cell cycle model. According to this model, cells go through a resting phase called G0, where they grow and prepare to enter the proliferative phase when they receive cell cycling checkpoint signals. Stem cells in the resting phase may enter the proliferative phase at a rate β, which incorporates negative feedback to signaling response pathways, or they can be removed from the resting pool at a rate κ through the biological processes such as differentiation, senescence, or death. The cells in the proliferating phase are randomly lost at a rate μ or undergo mitosis at a fixed time τ after entering the proliferative compartment. Each mother cell generated two daughter cells at mitosis. The newborn cells enter the resting phase to start the next cycle. This process is represented in Fig. 1.

During the resting phase, changes in cell numbers can be described using ordinary differential equations. However, during the proliferating phase, the biological processes can be modeled using the age-structured model, also known as the

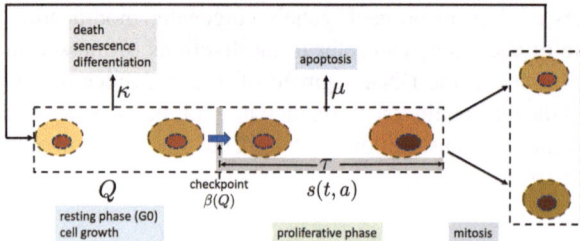

Fig. 1 The G0 model of stem cell regeneration. Here, Q represents the number of cells in the resting phase, and $s(t, a)$ represents the number of cells at time t with age a in the proliferating phase. Stem cells in the resting phase may enter the proliferative with a rate β or can be removed from the resting pool with a rate κ. The cells in the proliferating phase are randomly lost at a rate μ or undergo mitosis at a time τ after entering the proliferative compartment

transport equation introduced by M'kendrick in 1925 for medical problems [14]. The age-structured stem cell dynamics model given below was first proposed by Burns and Tannock in 1970 [15]. Later, in 1978, Michael Mackey applied this model to study the mechanism of periodic hematopoiesis [16].

We assume that all cells have identical micro-scale properties and are only interested in the time course of cell numbers. Let $Q(t)$ be the number of cells in the resting phase, and $s(t, a)$ be the number of stem cells at time t with age a in the proliferating phase. The above biological process can be described by the following age-structured equation [15–17]

$$\frac{\partial s(t, a)}{\partial t} + \frac{\partial s(t, a)}{\partial a} = -\mu s(t, a), \quad t > 0,\ 0 < a < \tau$$

$$\frac{dQ}{dt} = 2s(t, \tau) - (\beta(Q) + \kappa)Q, \quad t > 0. \tag{15}$$

The boundary condition at age $a = 0$ is given by

$$s(t, 0) = \beta(Q(t))Q(t). \tag{16}$$

Here, the proliferation rate of resting phase cells is represented by the function $\beta(Q)$, which depends on the number of cells in the resting phase.

In (15), κ signifies the rate at which cells irreversibly transition out of the resting phase. This transition may result from various biological processes, such as differentiation, senescence, or cell death. For conciseness, we will henceforth denote κ as the "differentiation rate."

The first equation in (15) can be integrated through the method of characteristic line, which results in a delay differential equation

$$\frac{dQ}{dt} = -(\beta(Q) + \kappa)Q + 2e^{-\mu\tau}\beta(Q_\tau)Q_\tau, \tag{17}$$

where $Q_\tau(t) = Q(t - \tau)$. This equation describes the general population dynamics of stem cell regeneration.

Here, the lag time τ is automatically introduced following the age-structured equation in (15), and τ represents the duration of the proliferation phase. If the lag time τ is omitted, (17) becomes an ODE model

$$\frac{dQ}{dt} = (2e^{-\mu} - 1)\beta(Q)Q - \kappa Q. \tag{18}$$

Here, we note that by omitting the delay, we cannot simply set $\tau = 0$ in (17), but replace the term $e^{-\mu\tau}$ with $e^{-\mu}$ that represents the survival probability of a cell after the proliferation phase.

Biologically, the self-renewal ability of a cell is intricately linked to microenvironmental conditions, such as growth factors and various types of cytokines, as well as intracellular signaling pathways [18–21]. Despite the complexity of signaling pathways, the phenomenological formation of Hill function dependence can be derived from simple assumptions regarding the interactions between signaling molecules and receptors [22, 23].

For instance, we assume that the niche secretes the positive growth factors and that the cells release growth factor inhibitors. Different types of cytokines bind to cell surface receptors to regulate cell behavior. Let [L] denote the concentration of ligands for growth factor inhibitor; [R] denote the density of free receptor; [R*] denotes the density of activated receptors, and Q denotes the stem cell number. The total number of receptors is

$$[R] + [R^*] = mQ. \tag{19}$$

At the equilibrium, we have the following equation

$$[R][L]^n = K[R^*], \tag{20}$$

where K is the equilibrium constant. We assume that the active receptors inhibit cell proliferation, and hence the proliferation rate β is proportional to the fraction of free receptors on a cell, i.e.,

$$\beta = \beta_0 \frac{[R]}{mQ}.$$

From Eqs. (19) and (20), we obtain the fraction of free receptors:

$$\frac{[R]}{mQ} = \frac{K}{K + [L]^n}.$$

When the ligands are secreted from stem cells and cleaned constantly, the ligand concentration is proportional to the cell number, resulting in $[L] = \sigma Q$. These calculations lead to the following Hill-type function of the proliferation rate:

$$\beta(Q) = \beta_0 \frac{\theta^n}{\theta^n + Q^n}, \tag{21}$$

where β_0 represents the maximum proliferation rate of normal cells, and $\theta = \sqrt[n]{K}/\sigma$ is a constant for the half-effective cell number.

Moreover, considering cancer cells that may exhibit uncontrolled cell growth, an extra factor β_1 can be introduced, leading to

$$\beta(Q) = \beta_0 \frac{\theta^n}{\theta^n + Q^n} + \beta_1, \tag{22}$$

where the positive parameter β_1 accounts for potential mutations in cancer cells that enable sustained proliferative signaling or evasion of growth suppressors, representing a hallmark of cancer [24]. Physical interactions can restrict cell growth within a more realistic model, causing β_1 to decrease to zero as the cell number Q becomes sufficiently large.

From (17), the steady state $Q(t) \equiv Q^*$ is given by the equation

$$- (\beta(Q^*) + \kappa)Q^* + 2e^{-\mu\tau}\beta(Q^*)Q^* = 0,$$

which yields either $Q^* = 0$, or

$$\beta(Q^*) = \frac{\kappa}{2e^{-\mu\tau} - 1}. \tag{23}$$

When $\beta(Q)$ is given by (22), the Eq. (17) has a unique positive steady state if and only if

$$\beta_0 > \frac{\kappa}{2e^{-\mu\tau} - 1} - \beta_1 > 0.$$

In particular, when

$$\beta_1 \geq \frac{\kappa}{2e^{-\mu\tau} - 1}, \tag{24}$$

zero solution $Q \equiv 0$ is the only steady state and is unstable.

Let $\bar{Q} = 1/Q$, and we have the equation for $\bar{Q}(t)$:

$$\frac{d\bar{Q}}{dt} = (\beta(\bar{Q}^{-1}) + \kappa)\bar{Q} + 2e^{-\mu\tau}\beta(\bar{Q}_\tau^{-1})(\bar{Q}/\bar{Q}_\tau)\bar{Q}.$$

When (24) is satisfied, the zero solution $\bar{Q} \equiv 0$ is stable. Consequently, all positive solutions of the original Eq. (17) approach infinity when $t \to \infty$,[1] indicating uncontrolled growth. Therefore, the inequality (24) summarizes a general condition for uncontrolled growth, i.e., malignant tumors. The inequality is satisfied when β_1 is increased, along with the decrease of μ and κ. Biologically, these conditions correspond to self-sufficiency in growth, insensitivity to antigrowth signals, evasion of apoptosis, and dysregulation in the differentiation (or senescence) pathways. These are well-known hallmarks of cancer [24]. Hence, a simple analysis for the homogeneous stem cell regeneration model (17) can reveal key hallmarks of cancer.

---

[1] A formal mathematical proof of this statement remains open.

4 Modeling Cellular Heterogeneity

The Eq. (14) explains the time course of a single cell without considering population dynamics. On the other hand, (17) describes population dynamics but doesn't provide microscale information of individual cells. To understand stem cell regeneration with cellular heterogeneity, we need to combine microscale fluctuation with macroscale population dynamics. This can be achieved by applying the mathematical modeling framework of either Lagrange or Euler coordinates, similar to fluid dynamics.

4.1 Lagrange Coordinate Modeling

Based on the Lagrange coordinate modeling, we consider each cell individually. Thus, let $\Sigma_t = \{[C_i]_{i=1}^{Q(t)}\}$ be the collection of all cells at time t, where $Q(t)$ represents the number of cells and $C_i = (\mathbf{x}_i(t), \mathbf{q}_i(t))$. We extend Eq. (14) to include all cells

$$
\begin{cases}
\dfrac{d\mathbf{x}_i}{dt} = \mathbf{F}(\mathbf{x}_i; \mathbf{q}_i), & \text{Between cell divisions} \\
(\mathbf{x}_i, \mathbf{q}_i) \mapsto (\mathbf{x}_i', \mathbf{q}_i') \sim \mathcal{P}(\mathbf{x}_i', \mathbf{q}_i' | \mathbf{x}_i, \mathbf{q}_i), & \text{Cell division}
\end{cases}
\tag{25}
$$

Here, each cell corresponds to a set of differential equations. The cell number $Q(t)$ changes with time due to cell death and cell division, so the total number of equations in (25) varies over time. Moreover, the cells do not divide synchronously in the Eq. (25); hence, it is challenging to write down a unified equation for all cells.

The Lagrange coordinate model (25) describes the microscale dynamics inside each cell; however, it is mathematically difficult to formulate and study. In numerical studies, we can apply the agent-based modeling technique to simulate the dynamics of multiple cells.

4.2 Euler Coordinate Modeling

To model cellular heterogeneity within the framework of Euler coordinate modeling, we introduce a variable $\mathbf{x}$ (often a high-dimensional vector) for the epigenetic state of a cell and Ω for the space of all possible epigenetic states in resting phase stem cells [23, 25, 26]. The epigenetic state $\mathbf{x}$ represents intrinsic cellular states that may dynamically change over time, whether in a cell cycle or during cell division. Biologically, the epigenetic state of a cell can be any molecular level changes that are independent of the DNA sequences, including the patterns of DNA methylation, nucleosome histone modifications, and transcriptomics [27–32].

Through the epigenetic state $\mathbf{x} \in \Omega$, let $Q(t, \mathbf{x})$ represent the number of cells at time t in the resting phase and with epigenetic state $\mathbf{x}$. Now, the total cell number is given by

$$Q(t) = \int_{\Omega} Q(t, \mathbf{x}) d\mathbf{x}. \tag{26}$$

The proliferation of each cell is regulated by the signaling pathways that are dependent on extracellular cytokines released by all cells in the niche and the epigenetic state $\mathbf{x}$ of the cell [22, 33, 34]. Let $\zeta(\mathbf{x})$ be the rate of cytokine secretion by a cell with state $\mathbf{x}$, and

$$c(t) = \int_{\Omega} Q(t, \mathbf{x}) \zeta(\mathbf{x}) d\mathbf{x} \tag{27}$$

represents the effective concentration of cytokines to regulate cell proliferation. Similar to (22), the proliferation rate β can be written as a function of cytokine concentration c and the epigenetic state $\mathbf{x}$, i.e.,

$$\beta(c, \mathbf{x}) = \beta_0(\mathbf{x}) \frac{\theta(\mathbf{x})^n}{\theta(\mathbf{x})^n + c^n} + \beta_1(\mathbf{x}). \tag{28}$$

Moreover, the apoptosis rate μ, the cell cycle duration τ, and the differentiation rate κ are dependent on the epigenetic state $\mathbf{x}$ and are denoted by $\mu(\mathbf{x})$, $\tau(\mathbf{x})$, and $\kappa(\mathbf{x})$, respectively. We assume that these rates depend solely on the state of each cell without considering the cell-to-cell interactions.

During cell division, a single mother cell splits into two daughter cells. However, the daughter cells may not share the same epigenetic state as the mother cell, resulting in cell plasticity during cellular regeneration. To account for this plasticity, we introduce an inheritance function (*aka* transition function), denoted by $p(\mathbf{x}, \mathbf{y})$, which represents the probability that a daughter cell with state $\mathbf{x}$ has originated from a mother cell with state $\mathbf{y}$ after cell division, i.e., the conditional probability density

$$p(\mathbf{x}, \mathbf{y}) = P(\text{state of daughter cell } = \mathbf{x} \mid \text{state of mother cell } = \mathbf{y}). \tag{29}$$

The inheritance function is used to consider cell plasticity during each cell cycle. It is obvious that

$$\int_{\Omega} p(\mathbf{x}, \mathbf{y}) d\mathbf{x} = 1$$

for any $\mathbf{y} \in \Omega$.

Now, similar to (15), when stem cell heterogeneity is included, we obtain the corresponding age-structured model equation

$$
\begin{aligned}
\nabla' s(t, a, \mathbf{x}) &= -\mu(\mathbf{x}) s(t, a, \mathbf{x}), \quad (0 < a < \tau(\mathbf{x})) \\
\frac{\partial Q(t, \mathbf{x})}{\partial t} &= 2 \int_\Omega s(t, \tau(\mathbf{y}), \mathbf{y}) p(\mathbf{x}, \mathbf{y}) d\mathbf{y} - (\beta(c(t), \mathbf{x}) + \kappa(\mathbf{x})) Q(t, \mathbf{x}),
\end{aligned} \tag{30}
$$

and

$$
s(t, 0, \mathbf{x}) = \beta(c(t), \mathbf{x}) Q(\mathbf{x}, t), \quad c(t) = \int_\Omega Q(t, \mathbf{x}) \zeta(\mathbf{x}) d\mathbf{x}.
$$

Here, $\nabla' = \partial/\partial t + \partial/\partial a$ represents the age-structured operator, and the epigenetic state $\mathbf{x}$ can be considered as parameters for the first equation. We can apply the characteristic line method to solve the first equation of (30) and obtain

$$
s(t, \tau(\mathbf{x}), \mathbf{x}) = \beta(c(t - \tau(\mathbf{x})), \mathbf{x}) Q(t - \tau(\mathbf{x}), \mathbf{x}) e^{-\mu(\mathbf{x})\tau(\mathbf{x})}.
$$

Thus, substituting $s(t, \tau(\mathbf{x}), \mathbf{x})$ into the second equation in (30), we obtain the following delay differential-integral equation (here, we only show the equation for $t \geq \tau$ that is important for the long-term behavior)

$$
\begin{cases}
\dfrac{\partial Q(t, \mathbf{x})}{\partial t} = -Q(t, \mathbf{x})(\beta(c, \mathbf{x}) + \kappa(\mathbf{x})) \\
\qquad\qquad + 2 \displaystyle\int_\Omega \beta(c(t - \tau(\mathbf{y})), \mathbf{y}) Q(t - \tau(\mathbf{y}), \mathbf{y}) e^{-\mu(\mathbf{y})\tau(\mathbf{y})} p(\mathbf{x}, \mathbf{y}) d\mathbf{y}, \\
c(t) = \displaystyle\int_\Omega Q(t, \mathbf{x}) \zeta(\mathbf{x}) d\mathbf{x}.
\end{cases} \tag{31}
$$

Equation (31) provides a general mathematical framework for modeling the dynamics of heterogeneous stem cell regeneration with the epigenetic transition.

From (31), and integrating both sides of the equation, we obtain

$$
\begin{aligned}
\frac{dQ}{dt} &= -\int_\Omega Q(t, \mathbf{x})(\beta(c, \mathbf{x}) + \kappa(\mathbf{x})) d\mathbf{x} \\
&\quad + 2 \int_\Omega \beta(c_{\tau(\mathbf{x})}, \mathbf{x}) Q(t - \tau(\mathbf{x}), \mathbf{x}) e^{-\mu(\mathbf{x})\tau(\mathbf{x})} d\mathbf{x}.
\end{aligned} \tag{32}
$$

If we omit the heterogeneity, all rate functions are independent of $\mathbf{x}$ and $c(t) = Q(t)$, we re-obtain the delay differential equation (17) for homogeneous stem cell regeneration.

Biologically, Eq. (31) connects different scale components: the gene expression values at the single-cell level ($\mathbf{x}$), the population dynamic properties ($\beta(c, \mathbf{x})$, $\kappa(\mathbf{x})$, and $\mu(\mathbf{x})$), cell cycle ($\tau(\mathbf{x})$), cytokine secretion ($\zeta(\mathbf{x})$), and the transition of epigenetic states ($p(\mathbf{x}, \mathbf{y})$). In this equation, the functions $\beta(c, \mathbf{x})$, $\kappa(\mathbf{x})$, $\mu(\mathbf{x})$, $\tau(\mathbf{x})$ describe the kinetic properties of cell cycling and are termed as the *kinetotype* of a cell [23]. This framework can be applied to different problems related to cell regeneration, such as development, aging, and tumor evolution [23, 26, 35, 36].

For mathematical simplicity, we can omit the delay τ in Eq. (31). It is important to note that the delay originates from the duration of the proliferative phase. Therefore, when the delay is omitted, we set $\tau(\mathbf{y}) = 0$ in $c(t - \tau(\mathbf{y}))$ and $Q(t - \tau(\mathbf{y}), \mathbf{y})$. Additionally, we replace $e^{-\mu(\mathbf{y})\tau(\mathbf{y})}$ with $e^{-\mu(\mathbf{y})}$ to represent the survival rate of cells during the proliferative phase. Consequently, when the delay is omitted, Eq. (31) becomes

$$
\begin{cases}
\dfrac{\partial Q(t, \mathbf{x})}{\partial t} = -Q(t, \mathbf{x})(\beta(c, \mathbf{x}) + \kappa(\mathbf{x})) \\
\qquad\qquad + 2 \displaystyle\int_{\Omega} \beta(c(t), \mathbf{y}) Q(t, \mathbf{y}) e^{-\mu(\mathbf{y})} p(\mathbf{x}, \mathbf{y}) d\mathbf{y}, \\
c(t) = \displaystyle\int_{\Omega} Q(t, \mathbf{x}) \zeta(\mathbf{x}) d\mathbf{x}.
\end{cases}
\tag{33}
$$

The discussions hereafter also apply to this equation when the delay is omitted.

The mathematical framework presented in Eq. (31) offers a basic model that incorporates the fundamental components of stem cell regeneration, such as cell cycling, cellular heterogeneity, and plasticity. However, stem cell systems in biological processes can be much more complex, and additional biological processes should be incorporated into the framework. For instance, the gene networks that underlie the dynamics of epigenetic states within a cell cycle, cell-to-cell interactions within a niche, and the interaction between all cells and the microenvironmental factors.

In the model given by (31), the variable $\mathbf{x}$ signifies the epigenetic state of a cell during the resting phase. For simplicity, we disregard changes in the epigenetic state within the resting phase. The model exclusively considers variations in the epigenetic state during cell division, encapsulated by the inheritance function $p(\mathbf{x}, \mathbf{y})$. However, this simplicity excludes the differentiation of cells during the resting phase, a departure from some biological observations indicating that differentiation can occur independently of cell division [37–42].

Several mathematical models have been proposed to describe division-independent differentiation, often formulated as a series of ordinary differential equations for discrete cell lineages or maturation stages [33, 43–45]. In our model, division-independent differentiation is represented by the differentiation rate $\kappa(\mathbf{x})$.

It is possible to extend the model Eq. (31) to include discrete cell lineages, resulting in the formulation

$$
\begin{aligned}
\frac{\partial Q_i(t, \mathbf{x})}{\partial t} &= \kappa_{i-1}(\mathbf{x}) Q_{i-1}(t, \mathbf{x}) - Q_i(t, \mathbf{x})(\beta_i(c_i, \mathbf{x}) + \kappa_i(\mathbf{x})) \\
&\quad + 2 \int_\Omega \beta_i(c_i(t - \tau_i(\mathbf{y})), \mathbf{y}) \\
&\qquad\qquad \times Q_i(t - \tau_i(\mathbf{y}), \mathbf{y}) e^{-\mu_i(\mathbf{y})\tau_i(\mathbf{y})} p_i(\mathbf{x}, \mathbf{y}) d\mathbf{y}, \\
c_i(t) &= \int_\Omega Q_i(t, \mathbf{x}) \zeta_i(\mathbf{x}) d\mathbf{x}.
\end{aligned}
\tag{34}
$$

Here, the subscript i denotes the i'th subtype of cells, along with their cell lineage or maturation stage. Mathematically, cell differentiation represented by κ is one-directional and irreversible, while cell plasticity, described by the inheritance function $p(\mathbf{x}, \mathbf{y})$, is multi-directional and reversible.

In a manner akin to the extension presented in (34), we can further extend the Eq. (31) to encompass gene mutations. Gene mutations typically occur in one direction following DNA replication during cell divisions. Consequently, we can introduce discrete mutant types and a mutation rate matrix $(p_{i,j}(\mathbf{x}))$, where $p_{i,j}$ denotes the mutation rate that changes the mutant type i to type j. The mutation rates may depend on the epigenetic state $\mathbf{x}$. The formulation involving gene mutation can be expressed as

$$
\begin{aligned}
\frac{\partial Q_i(t, \mathbf{x})}{\partial t} &= -Q_i(t, \mathbf{x})(\beta_i(c, \mathbf{x}) + \kappa_i(\mathbf{x})) \\
&\quad + 2 \int_\Omega (1 - \sum_{j \neq i} p_{i,j}(\mathbf{y})) \beta_i(c(t - \tau_i(\mathbf{y})), \mathbf{y}) \\
&\qquad\qquad \times Q_i(t - \tau_i(\mathbf{y}), \mathbf{y}) e^{-\mu_i(\mathbf{y})\tau_i(\mathbf{y})} p_i(\mathbf{x}, \mathbf{y}) d\mathbf{y}, \\
&\quad + 2 \sum_{j \neq i} \int_\Omega p_{j,i}(\mathbf{y}) \beta_j(c(t - \tau_j(\mathbf{y})), \mathbf{y}) \\
&\qquad\qquad \times Q_j(t - \tau_j(\mathbf{y}), \mathbf{y}) e^{-\mu_j(\mathbf{y})\tau_j(\mathbf{y})} p_j(\mathbf{x}, \mathbf{y}) d\mathbf{y} \\
c(t) &= \sum_i \int_\Omega Q_i(t, \mathbf{x}) \zeta_i(\mathbf{x}) d\mathbf{x}.
\end{aligned}
\tag{35}
$$

Please refer to [23] for examples and further discussions.

Please note that the term "stem cell" used in this context has a different meaning than its biological use. In organisms, stem cells are either undifferentiated or partially differentiated cells that can differentiate into various cell types and replicate indefinitely. Their self-renewal ability through cell division and their capacity to differentiate into specialized cell types distinguish stem cells from progenitor cells that cannot divide indefinitely and precursor or blast cells that are typically committed to differentiating into a specific cell type.

However, the mathematical models introduced here mainly focus on cell population dynamics without considering cell types. Therefore, the models do not explicitly include stem cell differentiation and cell type transitions. The model considers cells that can undergo proliferation, including stem cells, progenitor cells, or cancer cells. Each cell is either at the resting or proliferative phase during cell cycling. Cells that lose the proliferation ability are removed due to senescence, terminal differentiation, or death. Additionally, the specific cell types are not included in the model, as they are defined with specific patterns of marker gene expressions. The heterogeneity of cells is represented by the variance in the epigenetic states of cells, which are associated with the *kinetotype* of each cell rather than the cell type.

5 Modeling Cellular Plasticity

The inheritance function $p(\mathbf{x}, \mathbf{y})$ is vital to describe the plasticity of cells. However, the exact formula of the inheritance function is challenging to determine biologically, which is dependent on the complex biochemical reactions during the biological process of cell division. Nevertheless, while we consider $p(\mathbf{x}, \mathbf{y})$ as a conditional probability density, we can focus on the epigenetic state before and after cell division and omit the intermediate complex process.

Typically, our focus lies on the gene expressions that play an essential role in the gene network. The temporal dynamics of gene expression within one cell cycle can be described by a chemical rate equation of form (13). Cellular plasticity during cell division leads to the discontinuous changes in the cell states $\mathbf{x}$ and the parameters $\mathbf{q}$, and the long-term dynamics can be described by a discontinuous dynamical equation across cell division as (14). In this way, we can obtain phenomenological formulations for the inheritance function through numerical simulation based on gene regulatory networks and the laws of epigenetic state inheritance during cell division [46–48]. For more discussions, refer to [49].

Alternatively, we can assume the phenomenological formulations directly based on experimental observations. Let the epigenetic state $\mathbf{x} = (x_1, \cdots, x_m)$ represent m independent state variables, and assume that these states vary independently during cell division, we have

$$p(\mathbf{x}, \mathbf{y}) = \prod_{i=1}^{m} p_i(x_i, \mathbf{y}),$$

where $p_i(x_i, \mathbf{y})$ means the inheritance function of x_i, given the state $\mathbf{y}$ of the mother cells.

The inheritance function $p_i(x_i, \mathbf{y})$ can be represented as the density function of x_i, which is often phenomenological assumed based on the biological implications. For example, we can take the beta distribution for the normalized nucleosome modifications [46] or gamma distribution for transcription levels [50].

Here, we assume that $0 < x_i < 1$, and $p_i(x_i, \mathbf{y})$ is given by the density function of beta distribution, i.e.,

$$p_i(x_i, \mathbf{y}) = \frac{x_i^{a_i(\mathbf{y})-1}(1 - x_i)^{b_i(\mathbf{y})-1}}{B(a_i(\mathbf{y}), b_i(\mathbf{y}))}, \quad B(a, b) = \frac{\Gamma(a)\Gamma(b)}{\Gamma(a + b)}, \tag{36}$$

where $\Gamma(\cdot)$ represents the gamma function. Here, the inheritance function depends on two shape parameters, a and b, which are functions of the epigenetic state $\mathbf{y}$ of the mother cell. To determine the functions $a_i(\mathbf{y})$ and $b_i(\mathbf{y})$ from experimental data, we write the mean and variance of x_i, given the state $\mathbf{y}$, as

$$\mathrm{E}(x_i|\mathbf{y}) = \phi_i(\mathbf{y}), \quad \mathrm{Var}(x_i|\mathbf{y}) = \frac{1}{1 + \eta_i(\mathbf{y})}\phi_i(\mathbf{y})(1 - \phi_i(\mathbf{y})). \tag{37}$$

Accordingly, the shape parameters can be given by

$$a_i(\mathbf{y}) = \eta_i(\mathbf{y})\phi_i(\mathbf{y}), \quad b_i(\mathbf{y}) = \eta_i(\mathbf{y})(1 - \phi_i(\mathbf{y})) \tag{38}$$

through predefined functions $\phi_i(\mathbf{y})$ and $\eta_i(\mathbf{y})$. Here, the functions $\phi_i(\mathbf{y})$ and $\eta_i(\mathbf{y})$ should always satisfy

$$0 < \phi_i(\mathbf{y}) < 1, \quad \eta_i(\mathbf{y}) > 0.$$

The predefined functions, $\phi_i(\mathbf{y})$ and $\eta_i(\mathbf{y})$, are used to determine the conditional expectation and variance of the experimental data at the single cell level, as seen in Eq. (37). These functions act as a link between the model formulation and the experimental data. Additionally, $\phi_i(\mathbf{y})$ defines how the epigenetic state, x_i, depends on other components. This is often linked to the gene regulation network that underlies epigenetic states, and hence $\phi_i(\mathbf{y})$ can incorporate the information about gene networks.

A simple example of the above formulations can be shown by a single epigenetic state $x \in [0, 1]$, representing the stemness of a cell. Based on biological implications, cells with high stemness exhibit a low proliferation rate and an extremely low differentiation rate, the cells with intermediate stemness have a high proliferation rate and low differentiation rate, and the cells with low stemness have the lowest proliferation rate and the highest differentiation rate. Thus, we can formulate the proliferation rate β and the differentiation rate κ as (refer to [23, 35])

$$\beta(c, x) = \bar{\beta} \times \frac{\theta^n}{\theta^n + c^n} \times \frac{a_1 x + (a_2 x)^2}{1 + (a_3 x)^6}, \quad \kappa(x) = \kappa_0 \times \frac{1}{1 + (b_1 x)^6}, \tag{39}$$

where

$$c(t) = \int_0^1 Q(t, x)\zeta(x)\mathrm{d}x.$$

Here, we refer to Eq. (28) to define the proliferation rate β and assume $\beta_1(x) \equiv 0$. We should note that the mathematical form (39) is not unique. Any form describing the qualitative dependencies of β and κ on the stemness x would be acceptable.

Moreover, we assume that the apoptosis rate μ and the proliferating phase duration τ are independent of the stemness x.

The equation for heterogeneous stem cell regeneration becomes

$$
\frac{\partial Q(t, x)}{\partial t} = -Q(t, x)(\beta(c(t), x) + \kappa(x))
$$
$$
+ 2e^{-\mu\tau} \int_0^1 \beta(c(t - \tau), y)Q(t - \tau, y)p(x, y)dy,
\tag{40}
$$

where the inheritance function

$$
p(x, y) = \frac{x^{a(y)-1}(1 - x)^{b(y)-1}}{B(a(y), b(y))}, \quad B(a, b) = \frac{\Gamma(a)\Gamma(b)}{\Gamma(a + b)}.
$$

Equation (40) provides a simple example for mathematical studies. However, many fundamental problems remain open, such as the existence and uniqueness of the steady-state solution, the stability of steady-state solutions, and the existence of oscillatory solutions. For discussions of these problems, please refer to [26] or Sect. 9 in this chapter.

6 Individual-Based Model of Multicellular Tissues

Equation (31) provides a general mathematical framework to model stem cell regeneration when heterogeneity and plasticity of epigenetic or genetic states are included. This framework can describe many biological processes associated with stem cell regeneration, including development, aging, and cancer evolution [23]. Nevertheless, solving the equation numerically (31) is expensive when considering high-dimensional epigenetic states. Based on the above framework, we often develop hybrid computational models for multicellular tissues in applications.

Based on the mathematical framework (31), a hybrid numerical scheme was developed that combines a discrete stochastic process for the epigenetic/genetic states of individual cells with a continuous model of cell population growth [23]. In numerical simulation, a multicellular system is represented by a collection of multiple cells C_i, each cell has its epigenetic states $\mathbf{x}_i$, i.e., the system $\Omega_t = \left\{ [C_i(\mathbf{x}_i)]_{i=1}^{Q(t)} \right\}$, where $Q(t)$ represents the number of resting phase stem cells at time t. During a time interval $(t, t + \Delta t)$, each cell $(C_i(\mathbf{x}_i))$ undergoes cell fate decision, e.g., proliferation, apoptosis, or terminal differentiation, with a probability given by the kinetic rates. The probabilities of proliferation, apoptosis, or differentiation are given by $\beta(c, \mathbf{x}_i)\Delta t$, $\mu(\mathbf{x}_i)\Delta t$, or $\kappa(\mathbf{x}_i)\Delta t$, respectively, and hence are dependent on the epigenetic state of each cell as well as the microenvironmental condition $c =$

$\int Q(t, \mathbf{x})\zeta(\mathbf{x})d\mathbf{x}$. The total cell number $Q(t)$ changes after a time step Δt following the behaviors of all cells. When a cell undergoes proliferation, the epigenetic states of daughter cells change randomly according to the transition function $p(\mathbf{x}, \mathbf{y})$. In this hybrid model, all detailed molecular interactions are hidden within the kinetic rates and the inheritance function. Single-cell-based models can implement the proposed hybrid model through GPU architecture [51].

The above hybrid numerical scheme can also be integrated with the stochastic modeling of gene networks, through which the gene expression dynamics in individual cells are described with mathematical models of the form (25) between cell divisions. Each cell's kinetotype (β, μ, κ, and τ) depends on the gene expression state $\mathbf{x}$. For a detailed numerical scheme, refer to [52]. For more examples, please refer to [23, 35, 36, 52, 53].

7   Waddington Landscape

Understanding cell fate decisions and cell differentiation is crucial in biology. Waddington's epigenetic landscape is a fundamental concept in this understanding [4]. The concept visualizes a cell as a ball rolling on a mountain, where valleys correspond to stable cell phenotypes and ridges represent cell fate choices leading to new phenotypes. In other words, the landscape analogy illustrates how a cell's gene expression and environment interact to determine its development into a specific cell type. Mathematically, the definition of the Waddington landscape and its evolution are crucial in telling us more than reasoning about the process of cell fate decision during tissue development.

The Waddington landscape refers to the probability of a cell choosing a specific phenotype while developing. Thus, we can define the Waddington landscape as akin to the energy landscape in descriptions of physical systems, and the cell state probability is analogous to the Boltzmann distribution in statistical mechanics. Thus, we define the Waddington landscape as a potential $U(\mathbf{x})$, where $\mathbf{x}$ represents the state or phenotype of a cell. The likelihood of a cell in a state $\mathbf{x}$ is given by

$$P(\mathbf{x}) = Ce^{-\gamma U(\mathbf{x})}$$

according to the Boltzmann distribution, or the Arrhenius equation [54, 55], where γ is a constant, and C is a normalization coefficient. Conversely, the Waddington landscape is associated with the probability density $P(\mathbf{x})$ through (up to a scaling factor and constant translation)

$$U(\mathbf{x}) = -\log P(\mathbf{x}). \tag{41}$$

The potential $U(\mathbf{x})$ usually depends on the microenvironment conditions during tissue development and evolves with time. Hence, the evolutions of the population density and the Waddington landscape are represented by the time-dependent functions $P(t, \mathbf{x})$ and $U(t, \mathbf{x})$, respectively.

7.1 Gene Circuit Dynamics Approach

There are two methods to formulate the temporal evolution of the potential $U(t, \mathbf{x})$. The gene circuit dynamics (11) under a noisy fluctuating environment can be formulated as

$$\frac{d\mathbf{x}}{dt} = \mathbf{F}(\mathbf{x}) + \boldsymbol{\eta}, \tag{42}$$

where $\boldsymbol{\eta} = (\eta_1, \eta_2, \cdots, \eta_n)$ is a multi-dimensional Gaussian noise term. The correlation satisfies $\langle \eta_i(\mathbf{x}, t_1)\eta_j(\mathbf{x}, t_2) \rangle = 2D\delta_{i,j}\delta(t_1 - t_2)$, with D as the diffusion coefficient. Considering a system of a multi-cellular system where the gene expressions of all cells follow an initial distribution density $P(0, \mathbf{x})$, the evolution of the probability density $P(t, \mathbf{x})$ can be formulated through the Fokker-Planck equation

$$\frac{\partial P(t, \mathbf{x})}{\partial t} + \nabla \cdot \mathbf{J}(t, \mathbf{x}) = 0, \tag{43}$$

where $\nabla = (\frac{\partial}{\partial x_1}, \cdots, \frac{\partial}{\partial x_n})$, and $\mathbf{J}(t, \mathbf{x})$ is the probability flux defined as

$$\mathbf{J} = \mathbf{F}P - D\nabla P. \tag{44}$$

The probability flux consists of two components: the drift $\mathbf{F}P$, which quantifies the inclination of the cell state to move in a specific direction, and the diffusion term $D\nabla P$, accounting for the random fluctuation in the cell state. Moreover, $U(t, \mathbf{x}) = -\log P(t, \mathbf{x})$ gives the evolution of the Waddington landscape.

From the Eq. (43), at the stationary state, the divergence of the probability flux $\mathbf{J}_{ss}$ vanishes $\nabla \cdot \mathbf{J}_{ss} = 0$, yielding the steady-state probability density $P_{ss}(\mathbf{x})$. We note $P_{ss}(\mathbf{x}) = e^{-U_{ss}(\mathbf{x})}$ at the steady-state, where $U_{ss}(\mathbf{x})$ represents the steady-state potential. Thus, Eq. (44) results in a decomposition of the force [56, 57]

$$\mathbf{F}(\mathbf{x}) = -D\nabla U_{ss}(\mathbf{x}) + \mathbf{J}_{ss}(\mathbf{x})/P_{ss}(\mathbf{x}). \tag{45}$$

Equation (45) decomposes the force term of the gene circuit dynamics into two components. The first part is the potential gradient, connected to the steady-state probability by $U = -\ln P_{ss}$. The second part is the curl flux force, establishing a link between the divergence-free steady-state probability flux $\mathbf{J}_{ss}$ and the steady-state probability P_{ss}.

Determining the landscape U involves obtaining the steady-state solution of the Fokker-Planck equation (43). This equation describes changes in the probability density of the molecular states of a cell. However, it may not be suitable for describing tissue development because it does not account for biological processes such as cell division and death.

To incorporate the process of cell division and cell death, we replace $P(t, \mathbf{x})$ with the cell population density $f(t, \mathbf{x})$ and introduce a coefficient $R(\mathbf{x})$ to represent the birth-death rate (BDR) of a cell with state $\mathbf{x}$. Cell proliferation occurs when $R(\mathbf{x}) > 0$, while cell death occurs when $R(\mathbf{x}) < 0$. Following the discussion proposed by [11], consider a cell ω with gene expression $\mathbf{X}_t(\omega)$ starting from $\mathbf{Y}_0$ at $t = 0$. The corresponding weighted stochastic dynamics in the Itô sense can be written as

$$\begin{cases} d\mathbf{X}_t(\omega) = \mathbf{F}(\mathbf{X}_t(\omega))dt + \sqrt{2D}\,d\mathbf{W}_t(\omega), \\ \mathbf{X}_t(\omega)|_{t=0} = \mathbf{Y}_0(\omega), \\ d\rho_t(\omega) = R(\mathbf{X}_t(\omega))\rho_t(\omega)dt, \\ \rho_t(\omega)|_{t=0} = 1. \end{cases} \tag{46}$$

Here $\rho_t(\omega)$ is a time-varying weight for cell ω. The population density $f(t, \mathbf{x})$ is linked to the above equation as

$$f(t, \mathbf{x}) = \mathbb{E}\{\rho_t(\omega)\delta(\mathbf{x} - \mathbf{X}_t(\omega))\}, \tag{47}$$

where δ is the Dirac delta function, the expectation is taken over all possible trajectories ω.

The Eqs. (46) and (47) result in the following population balance equation (refer to [11, 58] for detailed discussions)

$$\frac{\partial f}{\partial t} = \nabla \cdot (D\nabla f) - \nabla \cdot (f\mathbf{F}) + Rf. \tag{48}$$

Given the BDR $R(\mathbf{x})$, Eq. (48) describes the time evolution of the population density. From (48), the following constraint

$$\int R(\mathbf{x}) f_{ss}(\mathbf{x})d\mathbf{x} = 0$$

for $R(\mathbf{x})$ is required to ensure a biologically meaningful steady-state population density $f_{ss}(\mathbf{x})$.

As discussed in [11], denote by $P_U(x)$ the steady-state population density with the known BDR $R(\mathbf{x})$ and by $P_0(\mathbf{x})$ the steady-state population density with $R(\mathbf{x}) \equiv 0$ for (48). Then, two energy landscapes can be constructed as

$$U_{ss}(\mathbf{x}) = -\log P_U(\mathbf{x}) \tag{49}$$

and

$$V_{ss}(\mathbf{x}) = -\log P_0(\mathbf{x})/P_U(\mathbf{x}). \tag{50}$$

The potential $U_{ss}(\mathbf{x})$ drives the system or cells to the steady distribution whose metastable basins indicate cell types. While $V_{ss}(\mathbf{x})$ quantifies the changes in the potential caused by the influence of cell proliferation and death, the values of V depict the pluripotency, and its negative gradient field describes the differentiation direction. Thus, we can consider $U_{ss}(\mathbf{x}) + V_{ss}(\mathbf{x})$ as the Waddington landscape from a stem cell state to a differentiated cell state [11].

Through the two potentials $U_{ss}(\mathbf{x})$ and $V_{ss}(\mathbf{x})$, we can decompose the force terms as

$$F(\mathbf{x}) = -D\nabla U_{ss}(\mathbf{x}) - D\nabla V_{ss}(\mathbf{x}) + \mathbf{G}(\mathbf{x}), \tag{51}$$

where $\mathbf{G}(\mathbf{x})$ is the curl part that describes the non-gradient nature of the considered dynamics.

Moreover, substituting the force decomposition (51) into (48), we have

$$\frac{\partial f}{\partial t} = \nabla \cdot (D\nabla f) - \nabla \cdot (f(-D\nabla(U_{ss}(\mathbf{x}) + V_{ss}(\mathbf{x})) + \mathbf{G}(\mathbf{x}))) + R(\mathbf{x})f$$
$$= \nabla \cdot (D(\nabla f + f\nabla(U_{ss}(\mathbf{x}) + V_{ss}(\mathbf{x})))) - \nabla \cdot (f\mathbf{G}(\mathbf{x})) + R(\mathbf{x})f.$$

Thus, we obtain the population balance equation of the population density $f(t, \mathbf{x})$:

$$\frac{\partial f}{\partial t} = \nabla \cdot (D(\nabla f + f\nabla(U_{ss}(\mathbf{x}) + V_{ss}(\mathbf{x})))) - \nabla \cdot (f\mathbf{G}(\mathbf{x})) + R(\mathbf{x})f. \tag{52}$$

The Eq. (52) establishes the connection between population density and the stationary state probability density and the flux. Through the general framework equations (48) or (52), various data-based methods have been proposed to identify cluster/cell types, differentiation trajectories, pseudotime, and cell pluripotency from the experimental data, such as the population balance analysis (PBA) [58–60] and landscape of differentiation dynamics (LDD) [11, 61].

Finally, it is important to point out that there are different ways to construct the potential landscape from a gene circuit dynamics of the form (42). Here, we mainly introduced Wang's potential landscape defined from the steady-state distribution associated with the Fokker-Planck equation (43) [56, 57]. Alternative definitions include the Freidlin-Wentzell quasi-potential from the large deviation theory [62], and Ao's potential through stochastic differentiation decomposition and a generalized Lyapunov function [63, 64]. A summary of different types of potential can be found in [65].

7.2 Stem Cell Regeneration Approach

The Eq. (48) mentioned above is derived from the Fokker-Planck approach, following the stochastic differentiation equation (42) for gene circuit dynamics. However,
these equations do not account for cell division and plasticity. On the other hand,
we can use Eq. (31), which explicitly includes cell division and plasticity, to derive
the evolution dynamics of cell population density for stem cell regeneration.

We note that

$$Q(t) = \int_{\Omega} Q(t, \mathbf{x}) d\mathbf{x} \tag{53}$$

represents the total cell number. The relative cell number with epigenetic state $\mathbf{x}$ is
given by

$$f(t, \mathbf{x}) = \frac{Q(t, \mathbf{x})}{Q(t)}. \tag{54}$$

It is easy to have

$$\int_{\Omega} f(t, \mathbf{x}) d\mathbf{x} \equiv 1, \forall t > 0. \tag{55}$$

The function $f(t, \mathbf{x})$ depicts the evolution of the probability density of epigenetic
states, and the Waddington landscape is given by

$$U(t, \mathbf{x}) = -\log f(t, \mathbf{x}). \tag{56}$$

From Eqs. (31), (32), (54), and (55), we can derive the evolution equation for
$f(t, \mathbf{x})$ as following.

$$
\begin{aligned}
\frac{\partial f(t, \mathbf{x})}{\partial t} &= -\frac{Q(t, \mathbf{x})}{Q(t)^2} \frac{dQ}{dt} + \frac{1}{Q(t)} \frac{\partial Q(t, \mathbf{x})}{\partial t} \\
&= -f(t, \mathbf{x}) \frac{1}{Q(t)} \left(-\int_{\Omega} Q(t, \mathbf{y})(\beta(c, \mathbf{y}) + \kappa(\mathbf{y})) d\mathbf{y} \right. \\
&\qquad\qquad \left. + 2 \int_{\Omega} \beta(c_{\tau(\mathbf{y})}, \mathbf{y}) Q(t - \tau(\mathbf{y}), \mathbf{y}) e^{-\mu(\mathbf{y})\tau(\mathbf{y})} d\mathbf{y} \right) \\
&\quad + \frac{1}{Q(t)} \left(-Q(t, \mathbf{x})(\beta(c, \mathbf{x}) + \kappa(\mathbf{x})) \right. \\
&\qquad\qquad \left. + 2 \int_{\Omega} \beta(c_{\tau(\mathbf{y})}, \mathbf{y}) Q(t - \tau(\mathbf{y}), \mathbf{y}) e^{-\mu(\mathbf{y})\tau(\mathbf{y})} p(\mathbf{x}, \mathbf{y}) d\mathbf{y} \right) \\
&= f(t, \mathbf{x}) \int_{\Omega} f(t, \mathbf{y})(\beta(c, \mathbf{y}) + \kappa(\mathbf{y})) d\mathbf{y} - f(t, \mathbf{x})(\beta(c, \mathbf{x}) + \kappa(\mathbf{x}))
\end{aligned}
$$

$$-\frac{2}{Q(t)} \int_{\Omega} \beta(c_{\tau(\mathbf{y})}, \mathbf{y}) Q(t - \tau(\mathbf{y}), \mathbf{y}) e^{-\mu(\mathbf{y})\tau(\mathbf{y})} f(t, \mathbf{x}) d\mathbf{y}$$

$$+\frac{2}{Q(t)} \int_{\Omega} \beta(c_{\tau(\mathbf{y})}, \mathbf{y}) Q(t - \tau(\mathbf{y}), \mathbf{y}) e^{-\mu(\mathbf{y})\tau(\mathbf{y})} p(\mathbf{x}, \mathbf{y}) d\mathbf{y}$$

$$=\frac{2}{Q(t)} \int_{\Omega} \beta(c_{\tau(\mathbf{y})}, \mathbf{y}) Q(t - \tau(\mathbf{y}), \mathbf{y}) e^{-\mu(\mathbf{y})\tau(\mathbf{y})} (p(\mathbf{x}, \mathbf{y}) - f(t, \mathbf{x})) d\mathbf{y}$$

$$- f(t, \mathbf{x}) \int_{\Omega} f(t, \mathbf{y})((\beta(c, \mathbf{x}) + \kappa(\mathbf{x})) - (\beta(c, \mathbf{y}) + \kappa(\mathbf{y}))) d\mathbf{y}.$$

Thus, we have the equation

$$\frac{\partial f(t, \mathbf{x})}{\partial t} = \frac{2}{Q(t)} \int_{\Omega} \beta(c_{\tau(\mathbf{y})}, \mathbf{y}) Q(t - \tau(\mathbf{y}), \mathbf{y}) e^{-\mu(\mathbf{y})\tau(\mathbf{y})} (p(\mathbf{x}, \mathbf{y}) - f(t, \mathbf{x})) d\mathbf{y}$$

$$- f(t, \mathbf{x}) \int_{\Omega} f(t, \mathbf{y}) ((\beta(c, \mathbf{x}) + \kappa(\mathbf{x})) - (\beta(c, \mathbf{y}) + \kappa(\mathbf{y}))) d\mathbf{y}.$$

$$(57)$$

Moreover, since

$$\frac{\partial U(t, \mathbf{x})}{\partial t} = -\frac{1}{f(t, \mathbf{x})} \frac{\partial f(t, \mathbf{x})}{\partial t},$$

we have

$$\frac{\partial U(t, \mathbf{x})}{\partial t} = -\frac{2}{Q(t, \mathbf{x})} \int_{\Omega} \beta(c_{\tau(\mathbf{y})}, \mathbf{y}) Q(t - \tau(\mathbf{y}), \mathbf{y}) e^{-\mu(\mathbf{y})\tau(\mathbf{y})} (p(\mathbf{x}, \mathbf{y}) - f(t, \mathbf{x})) d\mathbf{y}$$

$$+ \int_{\Omega} f(t, \mathbf{y}) ((\beta(c, \mathbf{x}) + \kappa(\mathbf{x})) - (\beta(c, \mathbf{y}) + \kappa(\mathbf{y}))) d\mathbf{y}.$$

$$(58)$$

The Eq. (58) gives the evolution of the Waddington landscape of a system of stem cell regeneration with cell heterogeneity and plasticity.

When the system reaches the equilibrium state so that $Q(t)$ and $f(t, \mathbf{x})$ are independent of the time t, we write

$$Q(t) = Q^*, f(t, \mathbf{x}) = f^*(\mathbf{x}), U(t, \mathbf{x}) = U^*(\mathbf{x}), c(t) = \int Q(t, \mathbf{x})\zeta(\mathbf{x})d\mathbf{x} = c^*.$$

The Eq. (57) becomes

$$2 \int_{\Omega} \beta(c^*, \mathbf{y}) e^{-\mu(\mathbf{y})\tau(\mathbf{y})} f^*(\mathbf{y})(p(\mathbf{x}, \mathbf{y}) - f^*(\mathbf{x})) d\mathbf{y}$$

$$- f^*(\mathbf{x}) \int_{\Omega} f^*(\mathbf{y}) ((\beta(c^*, \mathbf{x}) + \kappa(\mathbf{x})) - (\beta(c^*, \mathbf{y}) + \kappa(\mathbf{y}))) d\mathbf{y} = 0$$

$$(59)$$

at the equilibrium state. This gives an integral equation for the population density at the equilibrium state. Define a nonlinear operator $\mathcal{F}_c$ as

$$\mathcal{F}_c[f] = 2 \int_{\Omega} \beta(c, \mathbf{y}) e^{-\mu(\mathbf{y})\tau(\mathbf{y})} f(\mathbf{y})(p(\mathbf{x}, \mathbf{y}) - f(\mathbf{x})) d\mathbf{y}$$

$$- f(\mathbf{x}) \int_{\Omega} f(\mathbf{y}) \left((\beta(c, \mathbf{x}) + \kappa(\mathbf{x})) - (\beta(c, \mathbf{y}) + \kappa(\mathbf{y})) \right) d\mathbf{y},$$

the Eq. (59) gives a nonlinear eigenvalue problem

$$\mathcal{F}_c[f] = 0. \tag{60}$$

The equilibrium state density function $f^*(\mathbf{x})$ corresponds to the eigenfunction of the operator $\mathcal{F}_c$ with a positive eigenvalue c^*. Accordingly, the Waddington landscape at the equilibrium state is

$$U^*(\mathbf{x}) = -\log f^*(\mathbf{x}). \tag{61}$$

Thus, the above mathematical formulation provides a general method of calculating the evolution of Waddington's landscape during tissue growth.

7.3 Combining the Gene Circuit Dynamics with Stem Cell Regeneration

We have introduced two methods to formulate the population density $f(t, \mathbf{x})$. The gene circuit dynamics approach results in Eq. (48), where a birth-death rate $R(\mathbf{x})$ is introduced to account for the effect of cell birth and death. The stem cell generation approach considers the heterogeneity and plasticity of cells during cell divisions and explicitly formulates the biological processes of proliferation, differentiation, and apoptosis through the kinetotype of cells. These two approaches separately describe the mechanisms of cell type switches, driven by noise perturbations to the gene network or epigenetic changes during cell division. Here, we propose an integrated approach that combines both processes.

In Eq. (48), the birth-death rate $R(\mathbf{x})$ depends solely on the state of the cell, excluding cell-to-cell interactions. Additionally, cell plasticity associated with cell division is not considered in (48). However, the stem cell regeneration approach, as indicated by Eq. (57), suggests a growth operator $\mathcal{R}$ for the population density as follows:

$$\mathcal{R}[f] = \frac{2}{Q(t)} \int_\Omega \beta(c_{\tau(\mathbf{y})}, \mathbf{y}) Q(t - \tau(\mathbf{y}), \mathbf{y}) e^{-\mu(\mathbf{y})\tau(\mathbf{y})} (p(\mathbf{x}, \mathbf{y}) - f(t, \mathbf{x})) d\mathbf{y}$$

$$- f(t, \mathbf{x}) \int_\Omega f(t, \mathbf{y}) ((\beta(c, \mathbf{x}) + \kappa(\mathbf{x})) - (\beta(c, \mathbf{y}) + \kappa(\mathbf{y}))) d\mathbf{y}.$$

$$(62)$$

This operator takes into account the regulation of cell proliferation, differentiation, and apoptosis through both microenvironment conditions (via the factor c) and cellular states (via the epigenetic state $\mathbf{x}$). Moreover, cell plasticity during cell division is also involved through the inheritance function $p(\mathbf{x}, \mathbf{y})$.

Replacing the birth-death term Rf with the growth operator $\mathcal{R}[f]$, the population balance equation (52) can be expressed as

$$\frac{\partial f}{\partial t} = \nabla \cdot (D\nabla f) - \nabla \cdot (f\mathbf{F}) + \mathcal{R}[f]. \tag{63}$$

Thus, the Eq. (63) together with (31)–(32), provides an integrative mathematical model for the evolution of the population density $f(t, \mathbf{x})$. This equation combines the dynamics of the gene regulation network with heterogeneous stem cell regeneration. Accordingly, $U(t, \mathbf{x}) = -\log f(t, \mathbf{x})$ gives the evolution of Waddington's epigenetic landscape.

In particular, we consider a simple situation where the heterogeneities in the kinetotype of cells are omitted, i.e., the rate functions β, κ, μ, τ are independent of the epigenetic state $\mathbf{x}$. Moreover, we assumed that total cell number $Q(t)$ approaches a stable state $Q(t) = Q^*$, and the factor $c = \int Q(t, \mathbf{x})d\mathbf{x} = Q^*$. Thus, the growth factor becomes

$$\mathcal{R}[f] = 2\beta^* e^{-\mu\tau} \left(\int f(t - \tau, \mathbf{y}) p(\mathbf{x}, \mathbf{y}) d\mathbf{y} - f(t, \mathbf{x}) \right).$$

Here, $\beta^* = \beta(Q^*)$ represents the proliferation rate at the steady state given by (23). Thus, the Eq. (63) becomes

$$\frac{\partial f}{\partial t} = \nabla \cdot (D\nabla f) - \nabla \cdot (f\mathbf{F}) + 2\beta^* e^{-\mu\tau} \left(\int f(t - \tau, \mathbf{y}) p(\mathbf{x}, \mathbf{y}) d\mathbf{y} - f(t, \mathbf{x}) \right).$$

$$(64)$$

At the stationary state, the stationary densify $f^*(\mathbf{x})$ satisfies a nonlocal elliptic equation

$$\nabla \cdot (D\nabla f^*) - \nabla \cdot (f^*\mathbf{F}) = -2\beta^* e^{-\mu\tau} \left(\int f^*(\mathbf{y}) p(\mathbf{x}, \mathbf{y}) d\mathbf{y} - f^*(\mathbf{x}) \right). \tag{65}$$

The mathematical formulations of the Waddington landscape of stem cell regeneration with cell plasticity are given by Eqs. (64) and (65). Many basic mathematical problems associated with these equations remain unsolved.

8 Applications

The mathematical framework for stem cell regeneration presented in this chapter is versatile, allowing for its application in modeling the dynamics of multi-cellular system development, encompassing processes such as tissue development and tumor progression. Here, we introduce three instances where this modeling framework has been applied to investigate tumor progression and cell type transitions. For in-depth exploration and comprehensive details, please refer to the associated references.

8.1 Abnormal Growth Induced by Changes in the Microenvironmental Conditions

We applied the model framework to describe the abnormal growth process of tumor cells induced by changes in the microenvironmental conditions (refer to [53]). Consider a system of cells where the epigenetic state of each cell is denoted as $\mathbf{x} = (x_1, x_2)$, with x_1 representing the stemness of a cell and x_2 indicating the malignancy of a cell. Both x_1 and x_2 are normalized to the interval $[0, 1]$, defining the epigenetic state $\mathbf{x} \in \Omega = [0, 1] \times [0, 1]$.

Referring to the expressions in (39), we assume that the proliferation rate β and differential rate κ depend on the stemness x_1, and are defined as

$$\beta(c, \mathbf{x}) = \beta_0(\mathbf{x}) \frac{1}{1 + (c/\theta)^n}, \quad \beta_0(\mathbf{x}) = \beta_0 \frac{a_1 x_1 + (a_2 x_1)^{s_1}}{1 + (a_3 x_1)^{s_1}} \tag{66}$$

and

$$\kappa(\mathbf{x}) = \frac{\kappa_0}{1 + (b_1 x_1)^{s_1}}. \tag{67}$$

Here, we take c as the cell number Q. The coefficient $\theta(\mathbf{x})$ represents the repression of cell proliferation through cell responses to micro-environmental cytokines and is dependent on the malignancy. Thus, we assume that θ increases with the malignancy x_2, leading to

$$\theta(\mathbf{x}) = \theta_0 + \theta_1 \frac{x_2^{s_2}}{\theta_2^{s_2} + x_2^{s_2}}, \tag{68}$$

where θ_0, θ_1, and θ_2 are parameters.

The microenvironmental condition may affect the fitness of a cell in a given environment. To consider this effect on cancer evolution, we introduced a microenvironment index u, representing the effects of the microenvironment on malignancy and cell survival. Assuming $0 < u < 1$, with larger u indicating a microenvironment more suitable for cells with higher malignancy, we can define the fitness of a cell as

$$g(u, x_2) = g_0 x_2^u (1 - x_2)^{1-u},$$

where g_0 is a constant. Moreover, we assume that the apoptosis rate μ is dependent on the microenvironmental index u and malignancy x_2, expressed as

$$\mu(u, x_2) = \frac{\mu_0}{1 + \rho e^{g(u, x_2)}}. \tag{69}$$

Here, μ_0 and ρ are constants, indicating a maximum apoptosis rate $\mu_0/(1+\rho)$ when the fitness $g = 0$. Better fitness implies a lower apoptosis rate of a cell.

Similar to the previous argument, the inheritance function

$$p(\mathbf{x}, \mathbf{y}) = p_1(x_1, \mathbf{y}) \times p(x_2, \mathbf{y}),$$

where $p_i(x_i, \mathbf{y})$ are density functions of beta distribution

$$p_i(x_i, \mathbf{y}) = \frac{x_i^{a_i(\mathbf{y})-1}(1 - x_i)^{b_i(\mathbf{y})-1}}{B(a_i(\mathbf{y}), b_i(\mathbf{y}))}, \quad B(a, b) = \frac{\Gamma(a)\Gamma(b)}{\Gamma(a + b)}.$$

The shape parameters $a_i(\mathbf{y})$ and $b_i(\mathbf{y})$ are defined by the predefined functions $\phi_i(\mathbf{y})$ and $\eta_i(\mathbf{y})$ according to (37) and (38). Specifically, we take $\eta_1(\mathbf{y}) = \eta_2(\mathbf{y}) = \eta$ as constants and let $\phi_1(\mathbf{y}) = \phi_1(y_1)$ and $\phi_2(\mathbf{y}) = \phi_2(y_2)$ be defined as

$$\phi_1(y_1) = c_1 + d_1 \times \frac{(\alpha_1 y_1)^{1.5}}{1 + (\alpha_1 y_1)^{1.5}}, \tag{70}$$

$$\phi_2(y_2) = c_2 + d_2 \times \frac{(\alpha_2 y_2)^{2.1}}{1 + (\alpha_2 y_2)^{2.1}}. \tag{71}$$

Here, c_1, c_2, d_1 are constants, and d_2 may depend on the micro-environmental index u. When the microenvironment becomes abnormal (increases of u), the cells tend to be more malignant, so d_2 increases with u.

An individual-based modeling based on the above formulations can reveal abnormal cell growth dynamics when the micro-environmental index changes from a normal value ($u = 0.1$) to an abnormal value ($u = 0.9$). For detailed discussions of the simulation results, refer to [53].

8.2 Cell Plasticity Induced Immune Escape After CAR-T Therapy

Cancer immunotherapy has marked a significant breakthrough in recent years. However, immune escape frequently occurs following immunotherapy administration [66–68]. Here, we present an example illustrating how the mathematical framework detailed above can describe cancer cell plasticity-induced immune escape after chimeric antigen receptor (CAR) T cell therapy. For a comprehensive discussion, please refer to [35].

CAR-T therapy targeting CD19 has proven effective against B-cell acute lymphoblastic leukemia (B-ALL). While many patients achieve complete response with a single infusion of CD19-targeted CAR-T cells, many experience relapse after therapy [69, 70]. Our recent experiment showed that relapsed tumors in mice after infusion with CD19-28z-T cells maintain CD19 expression but exhibit a subpopulation of $CD19^+CD34^+$ and $CD123^+CD34^+$ tumor cells, absent in control NGFT-28z-treated mice [35]. Based on this observation, we proposed key assumptions that CAR-T-induced tumor cells transition into hematopoietic stem-like cells (by promoting CD34 expression) and myeloid-like cells (by promoting CD123 expression), thereby evading CAR-T cell targeting [35].

According to these assumptions, each cell's epigenetic state is represented by marker genes CD19, CD22, CD34, and CD123, pivotal in CD19 CAR-T cell responses and cell lineage dynamics. The proliferation rate β and differentiation rate κ depend on CD34 expression, a marker of stemness, as follows:

$$\beta = \beta_0 \frac{\theta}{\theta + N} \times \frac{5.8[CD34] + (2.2[CD34])^6}{1 + (3.75[CD34])^6},$$

$$\kappa = \kappa_0 \frac{1}{1 + (4.0[CD34])^6}.$$

Here, N represents the total cell number. The apoptosis rate μ comprises a basal rate μ_0 and a rate associated with the CAR-T signal:

$$\mu = \mu_0 + \mu_1 \times \text{Signal},$$

where Signal represents the CAR-T signal and is defined as

$$\text{Signal} = f([CD34], [CD123]) \frac{\gamma_{19}[CD19]}{1 + \gamma_{19}[CD19] + \gamma_{22}[CD22]} R(t),$$

$$f([CD34], [CD123]) = \frac{1}{(1 + ([CD34]/X_0)^{n_0})(1 + ([CD123]/X_1)^{n_1})}.$$

Here, $R(t)$ is the predefined CAR-T activity. CD34 and CD123, markers of stem-like and myeloid-like cells, respectively, were assumed to inhibit CAR-T signaling.

Cell plasticity is defined similarly to the previous discussions. For instance, given the CD34 expression level in cycle k (denoted by u_k), the expression level for cycle $k + 1$ (represented by u_{k+1}) is a random number from a beta distribution with a probability density

$$P(u_{k+1} = u|u_k) = \frac{u^{a_{34}-1}(1-u)^{b_{34}-1}}{B(a,b)}, \quad B(a,b) = \frac{\Gamma(a)\Gamma(b)}{\Gamma(a+b)},$$

the shape parameters a and b depend on the conditional expectation and the conditional variance of u_{k+1}. When

$$\mathrm{E}(u_{k+1}|u_k) = \phi_{34}(u_k), \quad \mathrm{Var}(u_{k+1}|u_k) = \frac{1}{1+\eta_{34}}\phi_{34}(u_{34})(1-\phi_{34}(u_{34})),$$

then

$$a = \eta_{34}\phi_{34}(u_k), \quad b = \eta_{34}(1-\phi_{34}(u_k)).$$

We can assume η_{34} as a constant, and

$$\phi_{34}(u_k) = 0.08 + 1.06\frac{(\alpha_{34}u_k)^{2.2}}{1+(\alpha_{34}u_k)^{2.2}},$$

letting

$$\alpha_{34} = 1.45 + 0.16 \times [CD19] + A_{34} \times \text{Signal},$$

which represents the promotion of CD34 expression by CD19 and the CAR-T signal. For further details, please refer to [35].

Simulations presented in [35] effectively replicated experimental results and predicted that CAR-T cell-induced cell plasticity could lead to tumor relapse in B-ALL after CD19 CAR-T treatment.

8.3   Cell-Type Transition Mediated by Epigenetic Modifications

Understanding how adult stem cells delicately balance self-renewal and differentiation remains a crucial issue in biological science. Here, we introduce a hybrid model of stem cell regeneration based on the mathematical framework presented in this chapter. The model integrates a gene regulation network, epigenetic state inheritance, and cell regeneration, allowing for multi-scale dynamics from transcription regulation to cell population [52].

The hybrid model contemplates a multi-cellular system, with each cell possessing a gene regulation network involving two genes that self-activate and repress each other. Cell regeneration behavior is modeled using a G0 cell cycle model, and the stochastic inheritance of epigenetic states during cell division is represented through the inheritance function. For a comprehensive understanding of the model, please refer to [52].

Let x_1 and x_2 represent the expression levels of two genes, X_1 and X_2. The gene expression dynamics within one cell cycle can be modeled with ordinary differential equations:

$$\begin{cases} \dfrac{dx_1}{dt} = a_1(\rho_1 + (1 - \rho_1)\dfrac{x_1^n}{s_1^n + x_1^n}) + b_1\dfrac{s_2^n}{s_2^n + x_2^n} - k_1 x_1, \\[3mm] \dfrac{dx_2}{dt} = a_2(\rho_2 + (1 - \rho_2)\dfrac{x_2^n}{s_2^n + x_2^n}) + b_2\dfrac{s_1^n}{s_1^n + x_1^n} - k_2 x_2, \end{cases} \tag{72}$$

where a_1, a_2, ρ_1, ρ_2, s_1, s_2, n, b_1, b_2, k_1, and k_2 are non-negative parameters. Parameters a_1 and a_2 denote the maximum expression rates of self-activation of the two genes, while b_1 and b_2 are basal expression rates of the two genes without regression. Random fluctuations to the gene expression rates, for example, to a_1 and a_2, can be expressed as

$$a_i(\eta_i) = \alpha_i e^{\sigma_i \eta_i - \sigma_i^2/2}, \quad i = 1, 2,$$

where α_1 and α_2 are positive parameters for the average expression rates. Here, σ_1 and σ_2 represent the intensities of noise perturbations, and η_1 and η_2 are colored noise defined by Ornstein-Uhlenbeck processes:

$$d\eta_i = -(\eta_i/\zeta_i)dt + \sqrt{2/\zeta_i}dW_i(t), \quad i = 1, 2,$$

where $W_1(t)$ and $W_2(t)$ are independent Wiener process, and ζ_1 and ζ_2 are relaxation coefficients.

To incorporate the effects of epigenetic modification on gene regulation dynamics, it is known that epigenetic regulations, such as histone modification or DNA methylation, can interfere with chromatin structure that the expression levels a_1 and a_2 depend on the epigenetic modification states of the two genes, denoted by u_1 and u_2, respectively. The epigenetic states can refer to the fractions of marked nucleosomes or methylated CpG sites in a DNA segment of interest, and hence $\mathbf{u} = (u_1, u_2) \in \Omega = [0, 1] \times [0, 1]$. The epigenetic state primarily affects the chromatin structure and influences the chemical potential to initiate transcription. Thus, along with extrinsic noise perturbations, the expression rates a_1 and a_2 can be expressed as follows:

$$a_i(u_i, \eta_i) = \alpha_i e^{\lambda_i u_i} e^{\sigma_i \eta_i - \sigma_i^2/2}, i = 1, 2,$$

where $\alpha_i (i = 1, 2)$ represents the impact of the epigenetic modification states on expression levels.

The epigenetic states u_1 and u_2 undergo random changes only during cell division. Following the above argument, the random inheritance of the epigenetic state is represented by the inheritance function

$$p(\mathbf{u}, \mathbf{v}) = P(\text{state of daughter cell } = \mathbf{u} \mid \text{state of mother cell } = \mathbf{v}).$$

We assume that u_1 and u_2 vary independently during cell division, and hence,

$$p(\mathbf{u}, \mathbf{v}) = p_1(u_1, \mathbf{v}) p_2(u_2, \mathbf{v}),$$

where $p_i(u_i, \mathbf{v})$ represents the transition function of u_i, given the state $\mathbf{v}$ of the mother cell. Similar to the previous argument, we write the inheritance function $p_i(u_i, \mathbf{v})$ through the density function of beta-distribution, and assuming the conditional expectation and conditional variance of u_i (given the state $\mathbf{v}$) as:

$$E(u_i|\mathbf{v}) = \phi_i(\mathbf{v}), \quad \mathrm{Var}(u_i|\mathbf{v}) = \frac{1}{1 + \psi_i(\mathbf{v})} \phi_i(\mathbf{v})(1 - \phi_i(\mathbf{v})).$$

The function $\phi_i(\mathbf{v})$ and $\psi_i(\mathbf{v})$ together define the inheritance function. We assume that $\psi_i(\mathbf{v})$ remains constant, while $\phi_i(\mathbf{v})$ increases with v_i and is expressed by a Hill function as:

$$\psi_i(\mathbf{v}) = m_0, \phi_i(\mathbf{v}) = m_1 + m_2 \frac{(m_3 v_i)^{m_4}}{1 + (m_3 v_i)^{m_4}}, \quad \mathbf{v} = (v_1, v_2), i = 1, 2,$$

where m_j $(j = 0, 1, 2, 3, 4)$ are positive parameters.

The gene regulation network dynamics described above can be integrated with the G0 cell cycle model based on the dependence of kinetotypes on cell types.

The gene regulation network dynamics given by (72) with properly selected parameter values can exhibit different patterns of steady states. Accordingly, we define the phenotype of cells following gene expressions of the steady states as follows: stem cells (SC) have medium expressions in both X_1 and X_2, transit-amplifying cells 1 (TA1) have high expression in X_1 and low expression in X_2, transit-amplifying cells 2 (TA2) have low expression in X_1 and high expression in X_2, and transition cells (TC) otherwise.

In the G0 cell cycle model, we only consider cells with the ability to undergo cell cycling, and each cell has different proliferation and cell death rates dependent on its cell phenotype. Cells that have lost the ability to experience cell cycling are removed from the system.

Different cell types differ in their cell proliferation regulation. For SCs, the proliferation rate can be given by

$$\beta_{SC} = \beta_0 \frac{\theta}{\theta + Q(t)},$$

where $Q(t)$ represents the number of SC at time t, β_0 represents the maximum proliferation rate, and θ is a constant for the half-effective cell number. TA cells, however, are assumed to have the maximum proliferation rate so that

$$\beta_{TA1} = \beta_{TA2} = \beta_0.$$

The removal rate κ, the apoptosis rate μ, and the proliferation duration τ are assumed as constants. For details, please refer to [52].

Given the epigenetic state $\mathbf{u} = (u_1, u_2)$ of each cell, the gene expression state $\mathbf{x} = (x_1, x_2)$ dynamically evolves according to the differential equations (72). Accordingly, the cell phenotype and the kinetic rates β, κ, μ, and τ can change during a cell cycle. When a cell undergoes mitosis, the cell divides into two cells, and the epigenetic states of the two daughter cells are calculated based on the inheritance functions. Simulation results in [52] demonstrate that random inheritance of epigenetic states during cell divisions can spontaneously induce cell differentiation, dedifferentiation, and transdifferentiation. Moreover, interfering with epigenetic modifications and introducing additional transcription factors can alter the probabilities of dedifferentiation and transdifferentiation, revealing a mechanism underlying cell reprogramming.

9 Mathematical Problems

This chapter introduces a mathematical framework (31) of stem cell regeneration that considers cell heterogeneity and plasticity. This equation, a delay differential-integral equation, incorporates nonlocal transitions between different epigenetic states. Moreover, we further integrate stem cell regeneration dynamics with gene regulation network dynamics, leading to a novel population balance equation (63). This equation provides an integrative mathematical model describing the evolution of the population density $f(t, \mathbf{x})$, or Waddington's epigenetic landscape $U(t, \mathbf{x}) = -\log f(t, \mathbf{x})$. While these equations provide novel mathematical formulations for quantifying the biological process of heterogeneous stem cell regeneration, many basic mathematical problems in understanding the biological processes remain open.

9.1   Steady-State Solution

To consider the steady-state solution of (31), let $Q(t, \mathbf{x}) = Q(\mathbf{x})$ represent the steady state. The resulting equations are:

$$\begin{cases} 0 = -Q(\mathbf{x})(\beta(c, \mathbf{x}) + \kappa(\mathbf{x})) + 2\int_\Omega \beta(c, \mathbf{y})Q(\mathbf{y})e^{-\mu(\mathbf{y})\tau(\mathbf{y})}p(\mathbf{x}, \mathbf{y})d\mathbf{y}, \\ c = \int_\Omega Q(\mathbf{x})\varsigma(\mathbf{x})d\mathbf{x}. \end{cases}$$

(73)

Substituting c into the first equation, $Q(\mathbf{x})$ satisfies the nonlinear eigenvalue problem

$$\mathcal{L}_c[Q(\mathbf{x})] = 2\int_\Omega \beta(c, \mathbf{y})e^{-\mu(\mathbf{y})\tau(\mathbf{y})}p(\mathbf{x}, \mathbf{y})Q(\mathbf{y})d\mathbf{y} - (\beta(c, \mathbf{x}) + \kappa(\mathbf{x}))Q(\mathbf{x}) = 0.$$

(74)

Therefore, the problem of the existence and uniqueness of the steady-state solution reduces to finding a positive eigenvalue c of the operator $\mathcal{L}_c$ such that the corresponding eigenfunction $Q(\mathbf{x})$ is non-negative for all $x \in \Omega$. When the eigenvalue $c > 0$ exists, the solution of (73) can be obtained through rescaling the corresponding eigenfunction following the second equation in (73).

Identifying the existence and stability of steady states is crucial for understanding the persistence of different biological states. While specific cases have been studied, open questions persist for general scenarios.

In the case of a finite discrete epigenetic state, and when the proliferation rate β is independent of the epigenetic state, the steady-state problem has been discussed in [71]. Particularly, when β is independent of the epigenetic state, the eigenvalue problem was reformulated as:

$$\mathcal{A}[Q] = \frac{1}{\beta(c)}Q,$$

(75)

where $\mathcal{A}$ is a linear operator defined as

$$\mathcal{A}[Q] = \frac{1}{\kappa(\mathbf{x})}\left[2\int_\Omega e^{-\mu(\mathbf{y})\tau(\mathbf{y})}p(\mathbf{x}, \mathbf{y})Q(\mathbf{y})d\mathbf{y} - Q(\mathbf{x})\right].$$

(76)

Thus, $(\beta(c))^{-1}$ is a positive eigenvalue of the operator $\mathcal{A}$. The existence of the eigenvalue can be obtained following the Perron-Frobenius theorem when there are a finite number of discrete epigenetic states [71].

The study in [71] further discussed the uniqueness and stability of the steady state when the inheritance function $p(\mathbf{x}, \mathbf{y})$ is independent of the state of the mother cell, i.e., $p(\mathbf{x}, \mathbf{y}) = p(\mathbf{x})$.

In the case of a 1-dimensional epigenetic state variable, when the delay $\tau > 0$ in the original Eq. (31), it may have oscillation solutions, and the positive steady state solution, if it exists, is unstable [36]. Conditions for oscillation solutions were studied in [36] through numerical simulations. Interestingly, the plasticity of cells during cell division can alter the requirements for oscillation solutions. However, the mathematical basis for the conditions for an oscillatory solution in the population number $Q(t, \mathbf{x})$ is unknown.

Moreover, the nonlinear eigenvalue problem (74) can be rewritten as:

$$\mathcal{L}_c[Q(\mathbf{x})] = \gamma(c)Q(\mathbf{x}), \tag{77}$$

and look for the principle eigenvalue γ for any parameter $c > 0$. Here, the principle eigenvalue corresponds to γ with a positive eigenfunction. Thus, if there is $c > 0$ such that $\gamma(c) = 0$, the problem (74) has a positive solution. The principle spectral theory has been applied [72] to study the existence, uniqueness, and multiplicity of positive steady states to the equations, as well as the long-time behavior of the time-dependent solutions, when the proliferation rate β is independent of the epigenetic state $\mathbf{x}$, and the delay τ is omitted in the original Eq. (31). Various explicit formulas for threshold values for tissue development, degeneration, and abnormal growth were obtained through the principle eigenvalue. For detailed discussions, please refer to [72].

9.2 Entropy Problem

Based on the population density, the entropy of the multicellular system at time t is defined as

$$S(t) = -\int_{\Omega} f(t, \mathbf{x}) \log f(t, \mathbf{x})d\mathbf{x}. \tag{78}$$

Thus, the derivative of the entropy $E(t)$ is given by

$$\frac{dS}{dt} = -\int_{\Omega} \frac{\partial f(t, \mathbf{x})}{\partial t} (1 + \log f(t, \mathbf{x})) \, d\mathbf{x}. \tag{79}$$

The entropy of a system measures the complexity of heterogeneity in the multicellular system, making the evolution of entropy biologically and mathematically interesting.

A study on abnormal growth induced by variations in microenvironmental conditions [52] revealed a nonlinear dependence of entropy changes on the cell population number $Q(t)$. As both $Q(t)$ and $S(t)$ are macroscopic measurements of the system, it is interesting to explore how we can formulate the macroscopic dynamics of the system during tissue growth.

Following [73, 74], the derivative of the entropy for the stochastic differential equation (42) is formulated as

$$D\frac{dS(t)}{dt} = e_p(t) - h_d(t), \tag{80}$$

where $S(t) = -\int_\Omega P(t, \mathbf{x}) \log P(t, \mathbf{x}) d\mathbf{x}$ is the entropy of the probability density function $P(t, \mathbf{x})$ at time t. Here, e_p is the entropy production rate (EPR)

$$e_p(t) = \int_\Omega |\mathbf{F}(\mathbf{x}) - D\nabla \log P(t, \mathbf{x})|^2 P(t, \mathbf{x}) d\mathbf{x}, \tag{81}$$

and h_d is the heat dissipation rate (HDR)

$$h_d(t) = \int_\Omega \mathbf{F}(\mathbf{x}) \cdot \mathbf{J}(t, \mathbf{x}) d\mathbf{x}, \tag{82}$$

with the probability flux $\mathbf{J}(t, \mathbf{x})$ define by (44). These formulas provide an interpretation of the gene circuit dynamics through statistical physics.

Now, for the entropy (78) associated with the population density $f(t, \mathbf{x})$, how can we formula the EPR and HDR? This question is crucial for understanding the statistical basis of tissue development as a non-equilibrium process, especially in the context of understanding entropy evolution during cancer development [75, 76].

9.3   Data-Driven Problem

In the model framework (31), while the kinetic rates β, κ, μ, and the proliferative duration τ can be obtained from cell regeneration dynamics, obtaining the inheritance function $p(\mathbf{x}, \mathbf{y})$ remains challenging with experimental data. Recent advancements in single-cell sequencing techniques have allowed quantification of gene expressions in individual cells. However, tracing cell division and quantifying molecular-level dynamics *in vivo* remain challenging. Novel techniques for tracing individual cell division and single-cell lineages have been developed [77, 78]. We anticipate that these methods may eventually enable the measurement of the inheritance function $p(\mathbf{x}, \mathbf{y})$.

Mathematically, addressing the following data-driven problem is crucial. Assuming we have obtained the population density $f(t, \mathbf{x})$ and the population size $Q(t)$ from experimental data, along with the knowledge of the kinetic rates β, κ, μ, and the duration τ, how can we derive the inheritance function $p(\mathbf{x}, \mathbf{y})$ based on the evolution equation (57)? Furthermore, how can we derive the force term $\mathbf{F}$ in the population balance equation (63)? Through these studies, we can gain insights into cell plasticity during cell regeneration, leading to an understanding of the dynamical Waddington landscape. A machine-learning method was proposed to construct a non-equilibrium potential landscape via a variational force projection formulation [79]. This method offers an approach to dealing with the high-dimensional potential landscape. Additionally, a method for quantifying the pluripotency landscape of cell differentiation from single-cell RNA sequencing (scRNA-seq) data from continuous birth-death process was developed in [61]. This study provides a computational tool to quantify cell potency within the Waddington landscape based on scRNA-seq data.

9.4 Local Epigenetic State Transition

In the Eqs. (31) or (33), the inheritance function $p(\mathbf{x}, \mathbf{y})$ represents the non-local transition of epigenetic states during cell division. Here, we consider the situation when only local transition is allowed, i.e.,

$$p(\mathbf{x}, \mathbf{y}) = \varphi(\mathbf{y} - \mathbf{x}) \tag{83}$$

so that $\varphi(\mathbf{z}) > 0$ only when $|\mathbf{z}| < \epsilon$ for some $\epsilon > 0$. The function $\varphi(\mathbf{z})$ satisfies

$$\int_{\mathbb{R}^m} \varphi(\mathbf{z}) d\mathbf{z} = 1, \quad \varphi(\mathbf{z}) \geq 0, \quad \forall \mathbf{z} \in \mathbb{R}^m. \tag{84}$$

We assume the epigenetic state $\mathbf{x} \in \Omega \subset \mathbb{R}^m$. We further let

$$\alpha_i = \int_{\mathbb{R}^m} z_i \varphi(\mathbf{z}) d\mathbf{z}, \quad D_{ij} = \int_{\mathbb{R}^m} z_i z_j \varphi(\mathbf{z}) d\mathbf{z}. \tag{85}$$

Now, we consider the Eq. (33), in which the delay τ is omitted. Substituting (83) into (33), we note that (here, we extend the integral over Ω to $\mathbb{R}^m$)

$$\int_{\mathbb{R}^m} \beta(c, \mathbf{y}) Q(t, \mathbf{y}) e^{-\mu(\mathbf{y})} p(\mathbf{x}, \mathbf{y}) d\mathbf{y}$$

$$= \int_{\mathbb{R}^m} \beta(c, \mathbf{x}) Q(t, \mathbf{x}) e^{-\mu(\mathbf{x})} \varphi(\mathbf{y} - \mathbf{x}) d\mathbf{y}$$

$$+ \int_{\mathbb{R}^m} (\mathbf{y} - \mathbf{x}) \cdot \nabla(\beta(c, \mathbf{x}) Q(t, \mathbf{x}) e^{-\mu(\mathbf{x})}) \varphi(\mathbf{y} - \mathbf{x}) d\mathbf{y}$$

$$+ \int_{\mathbb{R}^m} \frac{1}{2} (\mathbf{y} - \mathbf{x}) \cdot (\nabla^2(\beta(c, \mathbf{x}) Q(t, \mathbf{x}) e^{-\mu(\mathbf{x})}))(\mathbf{y} - \mathbf{x}) \varphi(\mathbf{y} - \mathbf{x}) d\mathbf{y}$$

$$= (1 + \sum_i \alpha_i \partial_i + \frac{1}{2} \sum_{i,j} D_{ij} \partial_{i,j}^2)(\beta(c, \mathbf{x}) Q(t, \mathbf{x}) e^{-\mu(\mathbf{x})}),$$

where

$$\partial_i = \frac{\partial}{\partial x_i}, \quad \partial_{i,j}^2 = \frac{\partial^2}{\partial x_i \partial x_j}.$$

Thus, the Eq. (33) is rewritten as

$$
\begin{cases}
\dfrac{\partial Q(t, \mathbf{x})}{\partial t} = (2 \sum_i \alpha_i \partial_i + \sum_{i,j} D_{ij} \partial_{i,j}^2)(\beta(c, \mathbf{x}) e^{-\mu(\mathbf{x})} Q(t, \mathbf{x})) \\[2mm]
\qquad\qquad + ((2e^{-\mu(\mathbf{x})} - 1)\beta(c, \mathbf{x}) - \kappa(\mathbf{x})) Q(t, \mathbf{x}) \qquad\qquad (86) \\[2mm]
c = \displaystyle\int_\Omega Q(t, \mathbf{x}) \zeta(\mathbf{x}) d\mathbf{x}.
\end{cases}
$$

Particularly, if $D_{ij} = D\delta_{ij}$, we have a reaction-diffusion equation

$$
\begin{cases}
\dfrac{\partial Q(t, \mathbf{x})}{\partial t} = \nabla \cdot (D\nabla + 2\boldsymbol{\alpha})(\beta(c, \mathbf{x}) e^{-\mu(\mathbf{x})} Q(t, \mathbf{x})) \\[2mm]
\qquad\qquad + ((2e^{-\mu(\mathbf{x})} - 1)\beta(c, \mathbf{x}) - \kappa(\mathbf{x})) Q(t, \mathbf{x}) \qquad\qquad (87) \\[2mm]
c = \displaystyle\int_\Omega Q(t, \mathbf{x}) \zeta(\mathbf{x}) d\mathbf{x},
\end{cases}
$$

where $\boldsymbol{\alpha} = (\alpha_1, \cdots, \alpha_m)$.

Equations (86) or (87) describe the dynamics of stem cell regeneration with the local transition of epigenetic states during cell division. These are linear evolution equations, with global regulation involved through the growth factor c. The problems of steady-state solution, entropy, and data-driven analysis mentioned above can be formulated based on the local epigenetic state transition.

10 Discussions

Cell division is a remarkable process in living organisms, where a parent cell divides into two daughter cells. These daughter cells can divide and grow independently, creating a new cell population from a single parental cell and its descendants. However, the two daughter cells may not be identical to their parental cell, leading to heterogeneity in the population of descendant cells due to the accumulation of epigenetic variations over multiple generations. The stable distribution of cell phenotypes forms the epigenetic landscape crucial in maintaining tissue home-ostasis despite random cell loss and regeneration. Quantitative modeling of the mechanism underlying the dynamic equilibrium of the epigenetic landscape during tissue development is of great significance in understanding the rules governing living organisms [80–83].

The characteristics of a cell are determined by how genes are expressed and regulated within the cell. Differentiation equations can be used to model the biochemical reactions of the gene network, either deterministic or stochastic. However, differential equation models only apply during a single cell cycle and become invalid when the cell divides. During cell division, both the cell states and the model parameters may change abruptly, leading to diversity and adaptability in stem cell regeneration over time. Therefore, it is vital to incorporate gene network dynamics with cell division when quantitatively modeling long-term biological processes like tissue development and tumor progression.

This chapter introduces two strategies for incorporating cell division into mathematical models of biological systems. The first strategy, Lagrange coordinate modeling, delves into the dynamics of gene networks within individual cells, resulting in random changes in cell states and model parameters during cell division. Each cell is represented by a unique set of equations, and the number of equations can change dynamically due to cell division. However, this approach makes the mathematical formulation challenging and lacks explicit inclusion of cellular behaviors such as proliferation, apoptosis, and differentiation in the model equations. To address these limitations, additional assumptions are necessary to describe the regulation of cell behaviors. Numerical studies of Lagrange coordinate models typically employ the agent-based modeling technique.

In contrast, the second strategy, Euler coordinate modeling, frames the evolution of population numbers of cells with the same epigenetic state through a differential-integral equation. This approach explicitly includes the regulations of cell behaviors, such as proliferation, apoptosis, and differentiation, in the equation. Plasticity resulting from cell division is represented as the inheritance probability function of epigenetic states, describing the conditional probability of epigenetic state changes in cell divisions. The inheritance function incorporates information on gene regulation networks. The Euler coordinate model integrates various biological interactions, encompassing the epigenetic state at the single-cell level, dynamic cell behavior, cytokine secretion, and the transition of epigenetic states. The concept of kinetotype is introduced through this model, offering a dynamic description of cell regeneration based on the epigenetic state of an individual cell [23]. Analogous to genotype, epigenotype, and phenotype, the concept of kinetotype provides a crucial description of cell type, which is often associated with the activities of specific genes enriched in related pathways. However, defining the kinetotype from single-cell sequencing data remains challenging.

These two modeling strategies are connected to different types of experimental data. Lagrange coordinate modeling describes molecular-level activities within individual cells, necessitating microscopic data for parameter identification. Such data include gene expression, epigenetic states, and protein levels associated with the genes of interest. On the other hand, Euler coordinate modeling focuses on the population-level dynamics of a multicellular system, requiring kinetic parameters related to cell renewal, differentiation, and cell death. Additionally, the heterogeneity and plasticity of cells are linked to single-cell sequencing data during the evolution process. Temporal single-cell RNA sequencing data can be crucial in determining how kinetic parameters may depend on the state of individual cells. However, the explicit dependencies are currently unknown and await further study.

The concept of Waddington's epigenetic landscape is crucial for deciphering the biological mechanisms that govern cell fate decisions and differentiation. This landscape is quantitatively characterized by the potential of epigenetic states, illustrated through the population density of cell states. Various modeling strategies exist to articulate the Waddington landscape, with one approach grounded in gene circuit dynamics featuring stochastic fluctuations in individual cells. This strategy leads to deriving the Fokker-Planck equation, aiming to model the evolution of population density.

However, the Fokker-Planck falls short in addressing critical biological processes, such as the regulation of proliferation, the impact of cell division, and cell plasticity induced by these divisions. These processes play pivotal roles in non-equilibrium biological phenomena like embryo development and tumorigenesis, thereby making the Fokker-Planck equation potentially misleading in capturing their intricacies.

Differing from the Fokker-Planck equation approach, the Euler coordinate model offers a more direct route to understanding the evolution of the Waddington landscape in the context of population dynamics during heterogeneous stem cell regeneration. This alternative approach derives an equation for the evolution of the population density of cells based on straightforward assumptions about cell proliferation, apoptosis, differentiation, and the transition of epigenetic states during cell division. In contrast to the Fokker-Planck approach, which delves into molecular-level dynamics inside a cell, the novel model equation (57) accentuates cellular-level dynamics described by the kinetotype of cells. Kinetotypes, representing fundamental properties of the non-equilibrium processes in tissue development and tumorigenesis [23, 24], provide a more rational comprehension of complex biological processes.

The Euler coordinate model has proven effective in studying diverse scenarios, including abnormal growth resulting from significant microenvironmental changes [53], the dynamic process of tumor cell plasticity-induced immune escape after CAR-T therapy in B-cell acute lymphoblastic leukemia (B-ALL) [35], and the dynamics of cell-type transitions mediated by epigenetic modifications [52]. In these scenarios, the evolution of cell population density induced by cell plasticity during division emerges as a critical determinant in achieving tissue homeostasis.

The Euler coordinate suggests a growth operator $\mathcal{R}$ for the population density, incorporating stem cell regeneration and plasticity during cell divisions. Applying the population balance analysis, we derive the dynamical equation (63) for the population density $f(t, \mathbf{x})$. This equation comprises three key components: the term $-\nabla(f\mathbf{F})$ captures the gradient of cell states influenced by gene regulation networks, the diffusion term $\nabla \cdot (D\nabla f)$ represents fluctuation random molecular perturbations, and the growth operator $\mathcal{R}[f]$ accounts for stem cell regeneration and plasticity. While cell types are defined by states that vanish the gradient force $\mathbf{F}$, the diffusion term elucidates potential cell type switches independent of cell division. In contrast, cell plasticity coupled with cell division is embodied by the inheritance function $p(\mathbf{x}, \mathbf{y})$ within the growth operator. Thus, Eq. (63) presents an integrated mathematical formulation harmonizing the dynamics of gene regulation networks with heterogeneous stem cell regeneration.

In the Euler coordinate model equation (31), there is an ongoing debate on how to measure the epigenetic state of a cell. While single-cell RNA sequencing techniques provide high-dimensional data for a single cell, a low-dimensional variable is still crucial in characterizing the cell type and its regeneration kinetics. Biologically, several marked genes linked to proliferation, differentiation, and apoptosis signal pathways are reasonable candidates for quantifying the epigenetic states. However, a subset of marker genes cannot represent the overall epigenetic state of a cell, and the dimension of marker gene expressions remains large.

Recently, machine learning methods have offered data-driven approaches to define macroscopic measurements of a cell based on single-cell RNA sequencing. These include the one-class logistic regression (OCLR) machine-learning method, which identifies the stemness feature [84], the Mann-Whitney U statistic method, which evaluates the signature score of genes associated with a signaling pathway [85], a computational framework (CytoTRACE) based on the number of expressed genes per cell [86], a computational framework (PhyloVelo) that estimates the velocity of transcriptomic dynamics by using monotonically expression genes through phylogenetic time [87], and the single-cell entropy (scEntropy), which measures the order of cellular transcriptome profile [88, 89]. These macroscopic measurements establish connections between the static single-cell RNA sequencing data and dynamic equations for the processes of tissue development. By integrating data-driven and model-driven studies, we can better understand complex biological processes, such as tissue development and tumor progression.

In summary, the Euler coordinate approach in the mathematical modeling of heterogeneous stem cell regeneration provides a reasonable understanding of the complex biological processes that govern the development of living organisms. The differential-integral equation model presented in this review offers a new perspective on the intricate problem of stem cell dynamics and a potential link between single-cell sequencing data and population dynamics.

Acknowledgments This work was supported by the National Natural Science Foundation of China (No. 12331018).

References

1. Eftimie, R., Gillard, J.J., Cantrell, D.A.: Mathematical models for immunology: current state of the art and future research directions. Bull. Math. Biol. **78**(10), 2091–2134 (2016)
2. Weinberg, R.A.: Using maths to tackle cancer. Nature **449**(7), 978–981 (2007)
3. Clevers, H.: What is an adult stem cell? Science **350**(6), 1319–1320 (2015)
4. Waddington, C.H.: The epigenotype. Int. J. Epidemiol. **41**(1), 10–13 (2012)

5. Ferrell, J.E.J.E.: Bistability, bifurcations, and Waddington's epigenetic landscape. Curr. Biol. **22**(11), R458–R466 (2012)
6. Hoppe, P.S., Schwarzfischer, M., Loeffler, D., Kokkaliaris, K.D., Hilsenbeck, O., Moritz, N., Endele, M., Filipczyk, A., Gambardella, A., Ahmed, N., Etzrodt, M., Coutu, D.L., Rieger, M.A., Marr, C., Strasser, M.K., Schauberger, B., Burtscher, I., Ermakova, O., Bürger, A., Lickert, H., Nerlov, C., Theis, F.J., Schroeder, T.: Early myeloid lineage choice is not initiated by random PU.1 to GATA1 protein ratios. Nature **535**(7611), 299–302 (2016)
7. Rommelfanger, M.K., MacLean, A.L.: A single-cell resolved cell-cell communication model explains lineage commitment in hematopoiesis. Development **148**(24), dev199779 (2021)
8. Fard, A.T., Srihari, S., Mar, J.C., Ragan, M.A.: Not just a colourful metaphor: modelling the landscape of cellular development using Hopfield networks. NPJ Syst. Biol. Appl. **2**, 16001 (2016)
9. Li, C., Wang, J.: Quantifying waddington landscapes and paths of non-adiabatic cell fate decisions for differentiation, reprogramming and transdifferentiation. J. R. Soc. Interface **10**(89), 20130787 (2013)
10. Feinberg, A.P., Levchenko, A.: Epigenetics as a mediator of plasticity in cancer. Science **379**(6632), eaaw3835 (2023)
11. Shi, J., Aihara, K., Li, T., Chen, L.: Energy landscape decomposition for cell differentiation with proliferation effect. Nat. Sci. Rev. **9**, nwac116 (2022)
12. Ferrell, J.E., Tsai, T.Y.C., Yang, Q.: Modeling the cell cycle: why do certain circuits oscillate? Cell **144**(6), 874–885 (2011)
13. Huh, D., Paulsson, J.: Random partitioning of molecules at cell division. Proc. Natl. Acad. Sci. USA **108**(36), 15004–15009 (2011)
14. M'Kendrick, L.C.A.G.: Applications of mathematics to medical problems. Proc. Edinburgh Math. Soc. **44**, 98–130 (1925)
15. Burns, F.J., Tannock, I.F.: On the existence of a G0-phase in the cell cycle. Cell Prolif. **3**(4), 321–334 (1970)
16. Mackey, M.C.: Unified hypothesis for the origin of aplastic anemia and periodic hematopoiesis. Blood **51**(5), 941–956 (1978)
17. Lei, J., Mackey, M.C.: Multistability in an age-structured model of hematopoiesis: cyclical neutropenia. J. Theor. Biol. **270**(1), 143–153 (2011)
18. Massague, J.: TGFβ signalling in context. Nat. Rev. Mol. Cell Biol. **13**(10), 616–630 (2012)
19. Moustakas, A., Pardali, K., Gaal, A., Heldin, C.H.: Mechanisms of TGF-β signaling in regulation of cell growth and differentiation. Immunol. Lett. **82**(1–2), 85–91 (2002)
20. Yang, L., Pang, Y.L., Moses, H.L.: TGF-β and immune cells: an important regulatory axis in the tumor microenvironment and progression. Trends Immunol. **31**(6), 220–227 (2010)
21. Ornitz, D.M., Itoh, N.: Fibroblast growth factors. Genome Biol. **2**(3), 3005.1–3005.12 (2001)
22. Bernard, S., Bélair, J., Mackey, M.C.: Oscillations in cyclical neutropenia: new evidence based on mathematical modeling. J. Theor. Biol. **223**(3), 283–298 (2003)
23. Lei, J.: A general mathematical framework for understanding the behavior of heterogeneous stem cell regeneration. J. Theor. Biol. **492**, 110196 (2020)
24. Hanahan, D., Weinberg, R.A.: The hallmarks of cancer. Cell **100**(1), 57–70 (2000)
25. Lei, J., Levin, S.A., Nie, Q.: Mathematical model of adult stem cell regeneration with cross-talk between genetic and epigenetic regulation. Proc. Natl. Acad. Sci. USA **111**(10), E880–E887 (2014)
26. Lei, J.: Evolutionary dynamics of cancer: from epigenetic regulation to cell population dynamics—mathematical model framework, applications, and open problems. Sci. China Math. **63**(3), 411–424 (2020)
27. Probst, A.V., Dunleavy, E., Almouzni, G.: Epigenetic inheritance during the cell cycle. Nat. Rev. Mol. Cell Biol. **10**(3), 192–206 (2009)
28. Schepeler, T., Page, M.E., Jensen, K.B.: Heterogeneity and plasticity of epidermal stem cells. Development **141**(13), 2559–2567 (2014)
29. Serra-Cardona, A., Zhang, Z.: Replication-coupled nucleosome assembly in the passage of epigenetic information and cell identity. Trends Biochem. Sci. **43**(2), 136–148 (2018)

30. Singer, Z.S., Yong, J., Tischler, J., Hackett, J.A., Altinok, A., Surani, M.A., Cai, L., Elowitz, M.B.: Dynamic heterogeneity and DNA methylation in embryonic stem cells. Mol. Cell **55**(2), 319–331 (2014)
31. Takaoka, K., Hamada, H.: Origin of cellular asymmetries in the pre-implantation mouse embryo: a hypothesis. Philos. Trans. R. Soc. Lond. B Biol. Sci. **369**, 20130536 (2014)
32. Wu, H., Zhang, Y.: Reversing DNA methylation: mechanisms, genomics, and biological functions. Cell **156**(1–2), 45–68 (2014)
33. Lander, A.D., Gokoffski, K.K., Wan, F.Y.M., Nie, Q., Calof, A.L.: Cell lineages and the logic of proliferative control. PLoS Biol. **7**(1), e15 (2009)
34. Mangel, M., Bonsall, M.B.: Stem cell biology is population biology: differentiation of hematopoietic multipotent progenitors to common lymphoid and myeloid progenitors. Theor. Biol. Med. Model. **10**(1), 5 (2012)
35. Zhang, C., Shao, C., Jiao, X., Bai, Y., Li, M., Shi, H., Lei, J., Zhong, X.: Individual cell-based modeling of tumor cell plasticity-induced immune escape after CAR-T therapy. Comput. Syst. Oncol. **1**, e21029 (2021)
36. Liang, X., Lei, J.: Oscillatory dynamics of heterogeneous stem cell regeneration. Commun. Appl. Math. Comput. **6**, 431–453 (2024)
37. O'Neill, M.C., Stockdale, F.E.: Differentiation without cell division in cultured skeletal muscle. Dev. Biol. **29**(4), 410–418 (1972)
38. Li, V.C., Kirschner, M.W.: Molecular ties between the cell cycle and differentiation in embryonic stem cells. Proc. Natl. Acad. Sci. USA **111**(26), 9503–9508 (2014)
39. Reilein, A., Melamed, D., Tavaré, S., Kalderon, D.: Division-independent differentiation mandates proliferative competition among stem cells. Proc. Natl. Acad. Sci. USA **115**(14), E3182–E3191 (2018)
40. Muhr, J., Hagey, D.W.: The cell cycle and differentiation as integrated processes: cyclins and CDKs reciprocally regulate Sox and Notch to balance stem cell maintenance. BioEssays **43**, 2000285 (2021)
41. Burda, I., Roeder, A.H.K.: Stepping on the molecular brake: slowing down proliferation to allow differentiation. Dev. Cell **57**, 561–563 (2022)
42. Kukreja, K., Patel, N., Megason, S.G., Klein, A.M.: Global decoupling of cell differentiation from cell division in early embryo. bioRxiv (2023). https://doi.org/10.1101/2023.07.29.551123
43. Adimy, M., Crauste, F., El Abdllaoui, A.: Asymptotic behavior of a discrete maturity stuctured system of hematopoietic stem cell dynamics with several delays. Math. Model. Nat. Phenom. **1**(2), 1–22 (2006)
44. Glauche, I., Cross, M., Loeffler, M., Roeder, I.: Lineage specification of hematopoietic stem cells: mathematical modeling and biological implications. Stem Cells **25**(7), 1791–1799 (2007)
45. Adimy, M., Angulo, O., Marquet, C., Sebaa, L.: A mathematical model of multistage hematopoietic cell lineages. Discrete Contin. Dyn. Syst. B **19**(1), 1–26 (2014)
46. Huang, R., Lei, J.: Dynamics of gene expression with positive feedback to histone modifications at bivalent domains. Int. J. Mod. Phys. B **4**, 1850075 (2017)
47. Huang, R., Lei, J.: Cell-type switches induced by stochastic histone modification inheritance. Discrete Contin. Dyn. Syst. B **24**, 5601–5619 (2019)
48. Sahoo, S., Mishra, A., Kaur, H., Hari, K., Muralidharan, S., Mandal, S., Jolly, M.K.: A mechanistic model captures the emergence and implications of non-genetic heterogeneity and reversible drug resistance in ER+ breast cancer cells. NAR Cancer **3**(3), zcab027 (2021)
49. Lei, J.: Systems Biology, Modeling, Analysis, and Simulation. Lecture Notes on Mathematical Modelling in the Life Sciences. Springer, Berlin (2021)
50. Cai, L., Friedman, N., Xie, X.S.: Stochastic protein expression in individual cells at the single molecule level. Nature **440**(7), 358–362 (2006)
51. Song, Y., Yang, S., Lei, J.: ParaCells: a GPU architecture for cell-centered models in computational biology. IEEE/ACM Trans. Comput. Biol. Bioinf. **16**(3), 994–1006 (2018)
52. Huang, R., Situ, Q., Lei, J.: Dynamics of cell-type transition mediated by epigenetic modifications. J. Theor. Biol. **577**, 111664 (2024)

53. Zhang, Y.D., Lei, J.: Entropy changes during the prolieferation of heteogeneous stem cells. Chin. J. Bininf. **22**(1), 67–78 (2024)
54. Hänggi, P., Talkner, P., Borkovec, M.: Reaction-rate theory: fifty years after Kramers. Rev. Mod. Phys. **62**(2), 251–342 (1990)
55. Zhou, J.X., Aliyu, M.D.S., Aurell, E., Huang, S.: Quasi-potential landscape in complex multi-stable systems. J. R. Soc. Interface **9**(77), 3539–3553 (2012)
56. Lapidus, S., Han, B., Wang, J.: Intrinsic noise, dissipation cost, and robustness of cellular networks: the underlying energy landscape of MAPK signal transduction. Proc. Natl. Acad. Sci. USA **105**(16), 6039–6044 (2008)
57. Wang, J., Zhang, K., Xu, L., Wang, E.: Quantifying the Waddington landscape and biological paths for development and differentiation. Proc. Natl. Acad. Sci. USA **108**(20), 8257–8262 (2011)
58. Weinreb, C., Wolock, S., Tusi, B.K., Socolovsky, M., Klein, A.M.: Fundamental limits on dynamic inference from single-cell snapshots. Proc. Natl. Acad. Sci. USA **115**(10), E2467–E2476 (2018)
59. Briggs, J.A., Weinreb, C., Wagner, D.E., Megason, S., Peshkin, L., Kirschner, M.W., Klein, A.M.: The dynamics of gene expression in vertebrate embryogenesis at single-cell resolution. Science **360**, 6392 (2018)
60. Fischer, D.S., Fiedler, A.K., Kernfeld, E.M., Genga, R.M.J., Bastidas-Ponce, A., Bakhti, M., Lickert, H., Hasenauer, J., Maehr, R., Theis, F.J.: Inferring population dynamics from single-cell RNA-sequencing time series data. Nat. Biotechnol. **37**, 461–468 (2019)
61. Shi, J., Li, T., Chen, L., Aihara, K.: Quantifying pluripotency landscape of cell differentiation from scRNA-seq data by continuous birth-death process. PLoS Comput. Biol. **15**(11), e1007488 (2019)
62. Freidlin, M.I., Wentzell, A.D.: Random Perturbation of Dynamical Systems. Springer, New York (1984)
63. Ao, P.: Potential in stochastic differential equations: novel construction. J. Phys. A Math. Gen. **37**, L25–L30 (2004)
64. Wang, H., Hu, W., Gan, X., Ao, P.: The generalized Lyapunov function as Ao's potential function: existence in dimensions 1 and 2. J. Appl. Anal. Comput. **13**(1), 359–375 (2023)
65. Zhou, P., Li, T.: Construction of the landscape for multi-stable systems: potential landscape, quasi-potential, A-type integral and beyond. J. Chem. Phys. **144**, 094109 (2016)
66. Kim, R., Emi, M., Tanabe, K.: Cancer immunoediting from immune surveillance to immune escape. Immunology **121**(1), 1–14 (2007)
67. Horn, L.A., Fousek, K., Palena, C.: Tumor plasticity and resistance to immunotherapy. Trends Cancer **6**(5), 432–441 (2020)
68. Hegde, P.S., Chen, D.S.: Top 10 challenges in cancer immunotherapy. Immunity **52**(1), 17–35 (2020)
69. Bhoj, V.G., Arhontoulis, D., Wertheim, G., Capobianchi, J., Callahan, C.A., Ellebrecht, C.T., Obstfeld, A.E., Lacey, S.F., Melenhorst, J.J., Nazimuddin, F., Hwang, W.T., Maude, S.L., Wasik, M.A., Bagg, A., Schuster, S., Feldman, M.D., Porter, D.L., Grupp, S.A., June, C.H., Milone, M.C.: Persistence of long-lived plasma cells and humoral immunity in individuals responding to CD19-directed CAR T-cell therapy. Blood **128**(3), 360–370 (2016)
70. Schuster, S.J., Svoboda, J., Chong, E.A., Nasta, S.D., Mato, A.R., Anak, Ö., Brogdon, J.L., Pruteanu-Malinici, I., Bhoj, V., Landsburg, D., Wasik, M., Levine, B.L., Lacey, S.F., Melenhorst, J.J., Porter, D.L., June, C.H.: Chimeric antigen receptor t cells in refractory b-cell lymphomas. N. Engl. J. Med. **377**(26), 2545–2554 (2017)
71. Situ, Q., Lei, J.: A mathematical model of stem cell regeneration with epigenetic state transitions. Math. Biosci. Eng. **14**(5–6), 1379–1397 (2017)
72. Su, Y.H., Li, W.T., Lou, Y., Wang, X.: Principal spectral theory for nonlocal systems and applications to stem cell regeneration models. J. Math. Pures Appl. **176**, 226–281 (2023)
73. Qian, H.: Mesoscopic nonequilibrium thermodynamics of single macromoleules and dynamic entropy-energy compensation. Phys. Rev. E **65**, 016102 (2001)

74. Zhang, X.J., Qian, H., Qian, M.: Stochastic theory of nonequilibrium steady states and its applications. Part I. Phys. Rep. **510**, 1–86 (2012)

75. Hanselmann, R.G., Welter, C.: Origin of cancer: an information, energy, and matter disease. Front. Cell Dev. Biol. **4**, 121 (2016)

76. Tarabichi, M., Antoniou, A., Saiselet, M., Pita, J.M., Andry, G., Dumont, J.E., Detours, V., Maenhaut, C.: Systems biology of cancer: entropy, disorder, and selection-driven evolution to independence, invasion and "swarm intelligence". Cancer Metastasis Rev. **32**(3–4), 403–421 (2013)

77. Quinn, J.J., Jones, M.G., Okimoto, R.A., Nanjo, S., Chan, M.M., Yosef, N., Bivona, T.G., Weissman, J.S.: Single-cell lineages reveal the rates, routes, and drivers of metastasis in cancer xenografts. Science **371**(6532), eabc1944 (2021)

78. Denoth-Lippuner, A., Jaeger, B.N., Liang, T., Royall, L.N., Chie, S.E., Buthey, K., Machado, D., Korobeynyk, V.I., Kruse, M., Munz, C.M., Gerbaulet, A., Simons, B.D., Jessberger, S.: Visualization of individual cell division history in complex tissues using iCOUNT. Cell Stem Cell **28**(11), 2020–2034 (2021)

79. Zhao, Y., Zhang, W., Li, T.: EPR-Net: constructing non-equilibrium potential landscape via a variational force projection formulation. Nat. Sci. Rev. **11**(7), nwae052 (2024)

80. Huang, S.: Non-genetic heterogeneity of cells in development: more than just noise. Development **136**(23), 3853–3862 (2009)

81. Brock, A., Chang, H., Huang, S.: Non-genetic heterogeneity–a mutation-independent driving force for the somatic evolution of tumours. Nat. Rev. Genet. **10**(5), 336–342 (2009)

82. Guillemin, A., Stumpf, M.P.H.: Non-equilibrium statistical physics, transitory epigenetic landscapes, and cell fate decision dynamics. Math. Biosci. Eng. **17**(6), 7916–7930 (2020)

83. Davila-Velderrain, J., Martinez-Garcia, J.C., Alvarez-Buylla, E.R.: Modeling the epigenetic attractors landscape: toward a post-genomic mechanistic understanding of development. Front. Genet. **6**, 160 (2015)

84. Malta, T.M., Sokolov, A., Gentles, A.J., Burzykowski, T., Poisson, L., Weinstein, J.N., Kamińska, B., Huelsken, J., Omberg, L., Gevaert, O., Colaprico, A., Czerwińska, P., Mazurek, S., Mishra, L., Heyn, H., Krasnitz, A., Godwin, A.K., Lazar, A.J., Cancer Genome Atlas Research Network, Stuart, J.M., Hoadley, K.A., Laird, P.W., Noushmehr, H., Wiznerowicz, M.: Machine learning identifies stemness features associated with oncogenic dedifferentiation. Cell **173**(2), 338–354 (2018)

85. Andreatta, M., Carmona, S.J.: UCell: robust and scalable single-cell gene signature scoring. Comput. Struct. Biotech. J. **19**, 3796–3798 (2021)

86. Gulati, G.S., Sikandar, S.S., Wesche, D.J., Manjunath, A., Bharadwaj, A., Berger, M.J., Ilagan, F., Kuo, A.H., Hsieh, R.W., Cai, S., Zabala, M., Scheeren, F.A., Lobo, N.A., Qian, D., Yu, F.B., Dirbas, F.M., Clarke, M.F., Newman, A.M.: Single-cell transcriptional diversity is a hallmark of developmental potential. Science **367**(6476), 405–411 (2020)

87. Wang, K., Hou, L., Wang, X., Zhai, X., Lu, Z., Zi, Z., Zhai, W., He, X., Curtis, C., Zhou, D., Hu, Z.: PhyloVelo enhances transcriptomic velocity field mapping using monotonically expressed genes. Nat. Biotechnol. **42**, 778–789 (2024)

88. Liu, J., Song, Y., Lei, J.: Single-cell entropy to quantify the cellular order parameter from single-cell RNA-seq data. Biophys. Rev. Lett. **15**(1), 35–49 (2020)

89. Ye, Y., Yang, Z., Zhu, M., Lei, J.: Using single-cell entropy to describe the dynamics of reprogramming and differentiation of induced pluripotent stem cells. Int. J. Mod. Phys. B **34**(30), 2050288 (2020)

A Mathematical Model to Describe the Formation of Perinuclear ATM Crown and the Effect of Irradiation and Antioxidants in Cells Affected by Alzheimer's Disease

Pauline Mazel, Nicolas Foray, and Laurent Pujo-Menjouet ⓘ

Abstract Alzheimer's disease is a neurodegenerative condition marked by the gradual and irreversible deterioration of brain cells. Normally, after oxidative stress, ATM proteins move to the nucleus to detect and repair DNA double-strand breaks. Recent research, however, reveals that in Alzheimer's-affected cells, ATM proteins form structures called perinuclear ATM crowns, influenced by the APOE protein, which prevents their migration to the nucleus.

We propose a mathematical model to describe perinuclear ATM crown formation using a system of nonlinear differential equations, aiming to assess the potential of irradiation and antioxidants as preventive and therapeutic interventions. Our model shows that irradiation or antioxidants alone cannot eliminate the perinuclear ATM crown, although antioxidants may play a preventive role if the crown has not yet formed. To dissolve an existing crown, a combined treatment with irradiation and antioxidants is necessary. Ultimately, this model serves as an initial framework for understanding disease progression and evaluating possible treatment strategies.

1 Introduction

Presently, dementia affects a global population exceeding 50 million individuals, with Alzheimer's disease (AD) representing the predominant form, as reported by the World Health Organization [18]. Those afflicted by AD commonly manifest

P. Mazel · L. Pujo-Menjouet (✉)
Universite Claude Bernard Lyon 1, CNRS, Ecole Centrale de Lyon, INSA Lyon, Université Jean Monnet, ICJ UMR5208, INRIA, Villeurbanne, France
e-mail: pauline.mazel@etu.univ-lyon1.fr; pujo@math.univ-lyon1.fr

N. Foray
Institut National de la Santé et de la Recherche Médicale, U1296 Research Unit "Radiation: Defense, Health, Environment", Centre Léon-Bérard, Lyon, France
e-mail: nicolas.foray@inserm.fr

Y. Mori et al. (eds.), *Dynamics of Physiological Control*, Lecture Notes on Mathematical Modelling in the Life Sciences,
https://doi.org/10.1007/978-3-031-82396-1_5

83

memory and behavioural disorders. Diagnostic challenges persist, and the onset of cognitive impairment often precedes formal diagnosis by several years. While a definitive cure is currently unavailable, early detection enhances the prospects for enhancing patient independence and quality of life [6, 15], mitigating the eventual, often fatal, consequences of the disease.

Two primary mechanisms of the disease are commonly under scrutiny. The first revolves around the β-amyloid peptide, which instigates the formation of amyloid deposits, while the second centres on the phosphorylation of the tau protein, culminating in neurofibrillary degeneration. Another protein implicated in the disease is the APOE protein (Apolipoprotein E), existing in various polymorphic forms: APOE $\varepsilon2$, $\varepsilon3$, and $\varepsilon4$. The $\varepsilon4$ allele's presence has been identified as an augmented risk factor for Alzheimer's disease (AD) development [17].

In this context, our specific focus is directed towards the role of the APOE protein (irrespective of its allele) as a potential interacting partner with the ATM kinase. The ATM kinase is a pivotal protein in the cellular response to genotoxic stress, playing a crucial role in the recognition and repair of DNA double-strand breaks (DSB), a pivotal cellular damage contributing to cell lethality [2, 3, 14].

1.1 Radiosensitivity

The exposure of cells to oxidative stress results in the production of DNA breaks in the nucleus, notably the DNA double-strand breaks (DSB) that are the key damage of cell lethality. In parallel, the same exposure contributes to dissociating multiprotein complexes localized in both the nucleus and cytoplasm. This is notably the case of the ATM, a key protein of the stress signalling that is predominantly localized as dimers (pATM) in the cytoplasm and dissociates in monomers in the linear proportion of the dose of radiation. Once dissociated, the two ATM monomers diffuse to the nucleus, cross the nuclear membrane, participate to the DSB recognition and trigger the DSB repair by recruiting some other proteins. This phenomenon, characterized by the migration of the ATM protein from the cytoplasm to the nucleus in response to radiation has been called Radiation-Induced NucleoShuttling of the ATM protein (RIANS) [2, 3]. The biological and clinical response to ionizing radiation depends on the rapidity and efficiency of the RIANS. Three groups of radiosensitivity have been defined to characterize the individual response to ionizing radiation of exposed individuals :

- **Group I: Radioresistance** (see Fig. 1a), cells from the individuals belonging to the group I show a very rapid RIANS in response to 2 Gy X-rays[1] (the dose of radiation per standard radiotherapy session). About 10 min after 2 Gy, all the radiation-induced DSB are recognized by ATM monomers.

---

[1] The gray (symbol: Gy) is the unit of ionizing radiation dose in the International System of Units (SI), defined as the absorption of one joule of radiation energy per kilogram of matter.

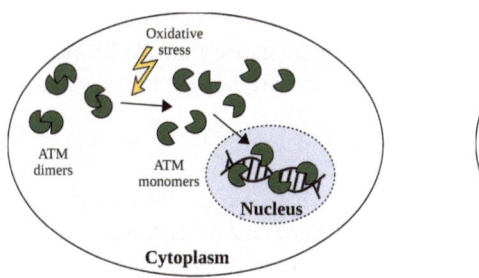 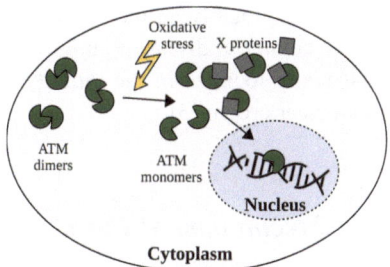

(a) Radioresistant cell. ATM monomers migrate into the nucleus to recognize and repair DSB.

(b) Radiosensitive cell. ATM monomers migrate into the nucleus, but a subset is stopped by the X proteins.

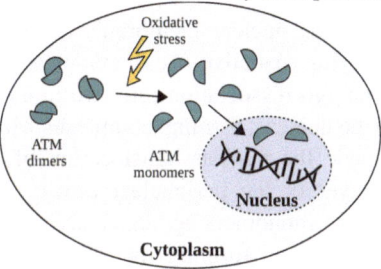

(c) Hyper-radiosensitive cell. Genetic mutations prevent the recognition and repair of DSB.

Fig. 1 Illustration of the three groups of radiosensitivity. (**a**) Radioresistant cell. (**b**) Radiosensitive cell. (**c**) Hyper-radiosensitive cell

- **Group II: Moderate radiosensitivity** (see Fig. 1b), cells from the individuals belonging to group II show very delayed RIANS in response to 2 Gy X-rays (the dose of radiation per standard radiotherapy session). Between 10 min and 1 h after 2 Gy X-rays, all the radiation-induced DSB are not recognized by ATM monomers. Before entering to the nucleus, some ATM monomers may associate with abnormally highly expressed ATM substrate proteins (called X proteins). The ATM-X complexes cannot diffuse to the nucleus and therefore cannot participate to the DSB recognition. As a result, some non-recognized DSB will be either non-repaired and lethal or managed by other impaired DSB repair pathways to produce misrepaired DSB (high risk of cancer) or else additional DSB (high risk of accelerated ageing).
- **Group III: Hyper-radiosensitivity** (see Fig. 1c), cells from the individuals belonging to group III show a lack of DSB recognition or a lack of DSB repair due to genetic mutations leading to the loss of the function of ATM protein or any other protein required in the DSB repair.

Alzheimer's disease is among the radiosensitive conditions to which the Radiation-Induced NucleoShuttling of the ATM protein (RIANS) model is applicable. Notably, in the context of fibroblasts derived from individuals afflicted by Alzheimer's disease, empirical evidence demonstrates that ATM proteins persistently localize in the vicinity of the nucleus, concretizing into a discernible perinuclear crown [3].

1.2 Mechanism of Formation of Perinuclear ATM Crown

The presence of the perinuclear ATM crown arises from a specific interaction between ATM proteins and a distinct subclass of X proteins exclusive to Alzheimer's disease (AD), namely the APOE protein. APOE proteins, typically localized in the cytoplasm and around the nuclear membrane, exhibit heightened expression levels in AD cells [3]. During oxidative stress, cytoplasmic ATM proteins in their dimeric configuration undergo dissociation into two monomers. These monomers subsequently migrate to the nucleus, forming complexes with APOE proteins. These complexes, in turn, impede the nuclear translocation of other ATM monomers, constituting the initial layer of the perinuclear crown. Progressing through the migration process, the ATM monomers aggregate around the nucleus, where they re-establish dimeric associations, forming the subsequent layers of the crown. Therefore, the perinuclear crown is characterized by a minimum of two layers: an inner layer comprising ATM-APOE complexes and an outer layer comprising ATM dimers (see Fig. 2).

In order to prevent or delay the formation of the perinuclear ATM crown and premature cell death, the effects of irradiation and antioxidants have been studied.

1.3 Irradiation and Antioxidants

Irradiation induces the dissociation of ATM dimers and ATM-APOE complexes within the perinuclear ATM crown. Notably, observations indicate the disappearance of phosphorylated ATM (pATM) signals subsequent to irradiation (10 minutes post-exposure to 2 Gy) [3]. However, this irradiation-induced dissociation does not culminate in the enduring elimination of the perinuclear crown.

Alternatively, the sustained presence of the crown can be mitigated by antioxidants. Their function involves the dispersion of ATM monomers, consequently impeding the re-dimerization process.[2] This dispersion of ATM monomers is anticipated to result in a substantially delayed, or potentially absent, reconstitution of the perinuclear crown.

---

[2] Data from ongoing research in N. Foray's laboratory, currently unpublished.

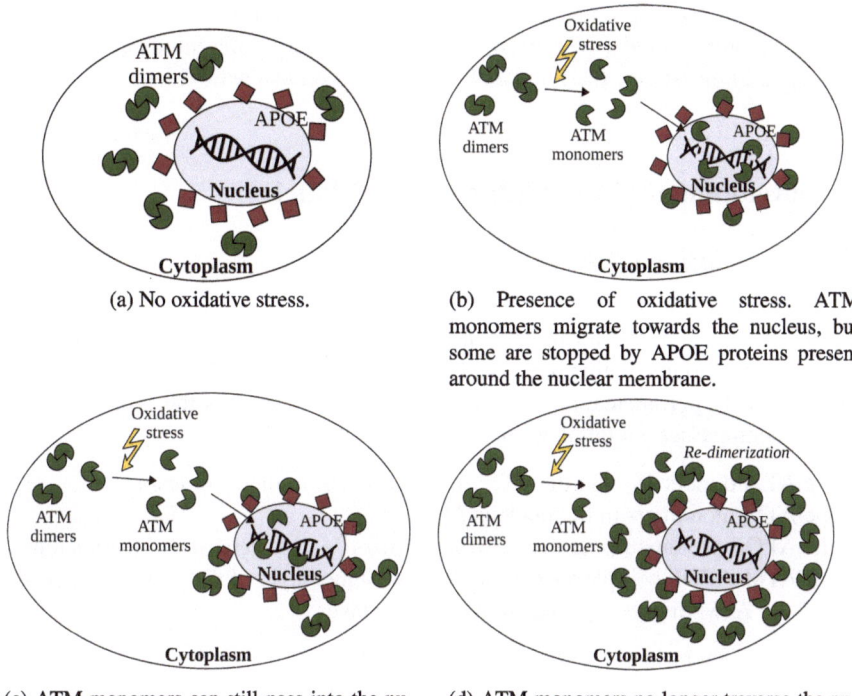

(a) No oxidative stress.

(b) Presence of oxidative stress. ATM monomers migrate towards the nucleus, but some are stopped by APOE proteins present around the nuclear membrane.

(c) ATM monomers can still pass into the nucleus, but a thin perinuclear crown begins to form.

(d) ATM monomers no longer traverse the nuclear membrane, they are blocked by the thick crown and re-dimerize. There is no longer recognition and repair of DSB.

Fig. 2 Evolution over time of a cell from a patient with Alzheimer's disease. (**a**) No oxidative stress. Presence of oxidative stress for (**b**), (**c**) and (**d**). (**b**) Beginning of migration. (**c**) Thin perinuclear crown. (**d**) Thicker perinuclear crown

1.4 State of the Art

In the past century, numerous mathematical models were introduced to elucidate clonogenic cell survival curves [5]. Early models were grounded in the target theory, with the seminal work by Crowther in 1924. Among these, the linear-quadratic (LQ) model, formulated in 1972 by Kellerer and Rossi [12, 13], has gained widespread adoption. Recently, a novel interpretation linking the Radiation-Induced NucleoShuttling of the ATM protein (RIANS) model to cell survival has been presented [4].

Concerning the specific context of Alzheimer's disease, our knowledge indicates the existence of a solitary mathematical model proposed to delineate the phenomenon [3]. In this work, we extend and generalize this model to furnish the most precise qualitative depiction of the underlying mechanism.

Hence, the primary objectives of this study are to formulate a model for the emergence of perinuclear ATM crowns utilizing ordinary differential equations and subsequently investigate the impact of both irradiation and antioxidants.

2 Development of the Mathematical Model

2.1 Description of the Model

The mathematical model formulated in this study adopts a compartmental structure, wherein each variable within the model is situated in distinct cellular compartments: the nucleus, the perinuclear region where the crown is formed, or the cytoplasm. The model encompasses a total of seven compartments (see Fig. 3):

- the ATM dimers D_C and the ATM monomers M_C in the cytoplasm,
- the ATM monomers in the nucleus M_N,
- the APOE proteins A and the ATM monomers M_A located around the nucleus,
- the ATM-APOE complexes C_A that form the inner layer of the perinuclear crown and the ATM dimers D_A that form the outer layer.

The monomers in the cytoplasm either re-dimerize (with constant $k_1 \in \mathbb{R}_+^*$) or migrate to the nucleus (function k_2). We assume that the perinuclear crown cannot indefinitely grow, and thus the function k_2 depends on D_A. We choose a

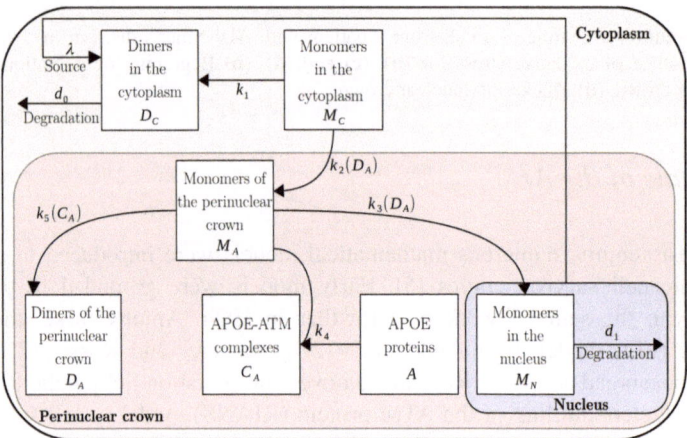

Fig. 3 diagram of protein interactions in the cell for our Alzheimer's disease model

Hill function (see Fig. 4a) defined for all $D_A \in \mathbb{R}_+$ by

$$k_2(D_A) = \frac{b_2 a_2^{n_2}}{a_2^{n_2} + D_A^{n_2}},$$

with $a_2, b_2, n_2 \in \mathbb{R}_+^*$.

Once the ATM monomers have migrated, they diffuse into the nucleus (function k_3), form the complexes of the first layer with APOE proteins (constant $k_4 \in \mathbb{R}_+^*$) or re-dimerize forming the second layer (function k_5). In this way, the diffusion into the nucleus depends on the thickness of the crown and thus depends on D_A. We take, for all $D_A \in \mathbb{R}_+$,

$$k_3(D_A) = \frac{b_3 a_3^{n_3}}{a_3^{n_3} + D_A^{n_3}},$$

with $a_3, b_3, n_3 \in \mathbb{R}_+^*$. This function has the same form as function k_2 (see Fig. 4a).

The re-dimerization of the ATM monomers forming the crown depends on the number of APOE-ATM complexes, that is to say, the quantity C_A (see Fig. 4b). We obtain for all $C_A \in \mathbb{R}_+$,

$$k_5(C_A) = \frac{a_5 C_A^{n_5}}{b_5^{n_5} + C_A^{n_5}},$$

with $a_5, b_5, n_5 \in \mathbb{R}_+^*$.

Note here that the Hill functions have often been used in biological models to describe saturation and molecular interactions (see [11] or [16] for instance).

We suppose that the production rate of ATM dimers is constant ($\lambda \in \mathbb{R}_+^*$) and its degradation rate $d_0 \in \mathbb{R}_+^*$ is proportional to the number of dimers. The ATM monomers in the nucleus have also a degradation rate $d_1 \in \mathbb{R}_+^*$.

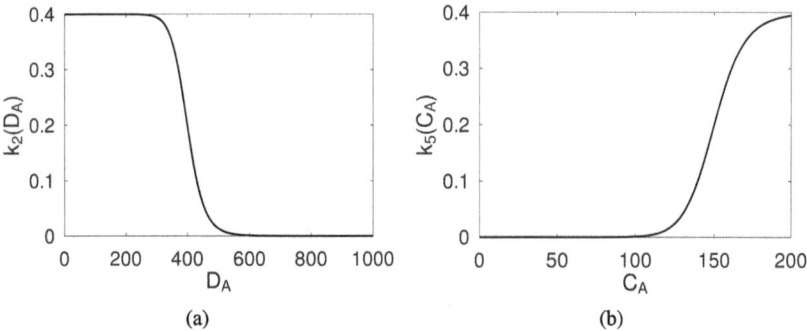

(a)   (b)

Fig. 4 Functions k_2 and k_5. (**a**) Function k_2 with $a_2 = 400$, $b_2 = 0.4$, $n_2 = 15$. (**b**) Function k_5 with $a_5 = 0.4$, $b_5 = 150$, $n_5 = 15$

Finally, we obtain the following system:

$$(S_0) \begin{cases} D'_C = \lambda - d_0 D_C + 0.5k_1 M_C^2, \\ M'_C = -k_1 M_C^2 - k_2(D_A)M_C, \\ M'_A = k_2(D_A)M_C - k_3(D_A)M_A - k_4 A M_A - k_5(C_A)M_A^2, \\ M'_N = k_3(D_A)M_A - d_1 M_N, \\ A' = -k_4 M_A A, \\ C'_A = k_4 M_A A, \\ D'_A = 0.5k_5(C_A)M_A^2. \end{cases}$$

It is important to emphasize here that the presence of the square in the M_C^2 term is a consequence of the dimerization process, which is delineated by the standard mass action law.

2.2 Existence, Uniqueness and Positivity

Proposition 1 (Existence, Uniqueness and Positivity) *Let* $(t_0, D_{C0}, M_{C0}, M_{N0}, M_{A0}, A_0, C_{A0}, D_{A0}) \in \mathbb{R} \times] - \min(a_2, a_3, a_5), +\infty[$[7].
Then there exists a unique solution for (S_0) *passing through*
$(t_0, D_{C0}, M_{C0}, M_{N0}, M_{A0}, A_0, C_{A0}, D_{A0})$. *Furthermore, if the initial conditions are positive, then the solution is positive.*

Proof Thanks to the Cauchy–Lipschitz theorem, we have the existence and uniqueness of the solutions.

Now, we assume all the initial conditions to be positive. Denote $t_1 \in \mathbb{R}$ the first time D_C is null. Then $D'_C(t_1) = \lambda + 0.5k_1 M_C^2(t_1)$ is positive, and thus D_C is increasing at t_1.

Similarly, denote t_1 the first time a quantity is null (t_1 can be different from the precedent one),

$$M_C(t_1) = 0 \implies M'_C(t_1) = 0,$$

$$M_A(t_1) = 0 \implies M'_A(t_1) = k_2(D_A(t_1))M_C(t_1),$$

$$M_N(t_1) = 0 \implies M'_N(t_1) = k_3(D_A(t_1))M_A(t_1),$$

$$A(t_1) = 0 \implies A'(t_1) = 0,$$

$$C_A(t_1) = 0 \implies C'_A(t_1) = k_4 M_A(t_1)A(t_1),$$

$$D_A(t_1) = 0 \implies D'_A(t_1) = 0.5k_5(C_A(t_1))M_A^2(t_1).$$

That allows us to conclude to the positivity of the different populations.

2.3 Study of Equilibria

We recall that the system was given by

$$\begin{cases} D'_C = \lambda - d_0 D_C + 0.5 k_1 M_C^2, \\ M'_C = -k_1 M_C^2 - k_2(D_A) M_C, \\ M'_A = k_2(D_A) M_C - k_3(D_A) M_A - k_4 A M_A - k_5(C_A) M_A^2, \\ M'_N = k_3(D_A) M_A - d_1 M_N, \\ A' = -k_4 M_A A, \\ C'_A = k_4 M_A A, \\ D'_A = 0.5 k_5(C_A) M_A^2. \end{cases}$$

We have $C'_A = -A'$, thus $C_A = C_{A0} + A_0 - A$, and we can simplify the system:

$$\begin{cases} D'_C = \lambda - d_0 D_C + 0.5 k_1 M_C^2, \\ M'_C = -k_1 M_C^2 - k_2(D_A) M_C, \\ M'_A = k_2(D_A) M_C - k_3(D_A) M_A - k_4 A M_A - k_5(C_{A0} + A_0 - A) M_A^2, \\ M'_N = k_3(D_A) M_A - d_1 M_N, \\ A' = -k_4 M_A A, \\ D'_A = 0.5 k_5(C_{A0} + A_0 - A) M_A^2. \end{cases}$$

The equilibria satisfy

$$\begin{cases} \lambda - d_0 D_C^* + 0.5 k_1 M_C^{*2} = 0, \\ -k_1 M_C^{*2} - k_2(D_A^*) M_C^* = 0, \\ k_2(D_A^*) M_C^* - k_3(D_A^*) M_A^* - k_4 A^* M_A^* - k_5(C_{A0} + A_0 - A^*) M_A^{*2} = 0, \\ k_3(D_A^*) M_A^* - d_1 M_N^* = 0, \\ k_4 M_A^* A^* = 0, \\ 0.5 k_5(C_{A0} + A_0 - A^*) M_A^{*2} = 0. \end{cases}$$

We have $k_4 M_A^* A^* = 0$, thus $M_A^* = 0$ or $A^* = 0$.

- If $M_A^* = 0$, then $M_C^* = 0$ thanks to the second equation and $M_N^* = 0$ thanks to the fourth equation. Finally, the first equation gives $D_C^* = \lambda/d_0$.
- If $M_A^* \neq 0$, then $A^* = 0$ and $A^* = C_{A0} + A_0$, which is impossible if we assume at least C_{A0} or A_0 strictly positive.

Finally, the equilibria are given by $(\lambda/d_0, 0, 0, A^*, C_{A0} + A_0 - A^*, D_A^*)$. In this way, we have an infinite number of equilibria that depend explicitly on the initial conditions A_0 and C_{A0}, but also implicitly on M_{C0} and D_{A0}.

3 Description of Perinuclear Crown Formation

Based on the work described in [3], we propose to modify the (S_0) system by inserting the oxidative stress effect in the model as follows.

3.1 Mathematical Modelling

The perinuclear crown forms gradually in response to daily stresses. According to the intensity of these stresses, it will contribute to the formation of the crown more or less quickly.

During stress, all the complexes and dimers dissociate in monomers, which is modelled by the function g (see Fig. 5). We propose to model these stresses in two ways:

- **Low successive stresses** representing the different stresses encountered in life, and modelled by a sum of Gaussian functions, defined for all $t \in \mathbb{R}_+$ by

$$g(t) = \sum_{i=1}^{N} c_s \exp\left(-\frac{(t - m_{sci})^2}{2\sigma_{sc}^2}\right),$$

where $N \in \mathbb{N}^*$ is the number of stresses, $c_s \in \mathbb{R}_+^*$ is the intensity of stress, $m_{sci} = iT$ for $i = 1, \ldots, N$ with $T \in \mathbb{R}_+^*$ the stress repetition period and $\sigma_{sc} \in \mathbb{R}_+^*$ determines the time interval between each repetition.

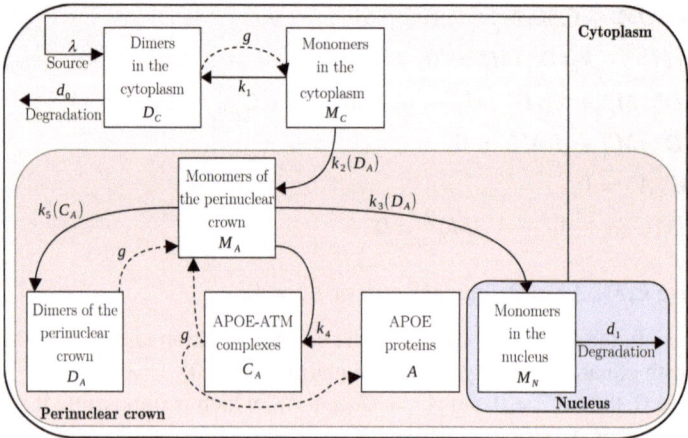

Fig. 5 Diagram of the model for Alzheimer's disease. The parameters with dashed arrows are those that depend on the stress

- A **constant and low stress** defined for all $t \in \mathbb{R}_+$ by

$$g(t) = c_s,$$

with $c_s \in \mathbb{R}_+^*$.

Finally, we obtain the following system:

$$(S) \begin{cases} D_C' = \lambda - d_0 D_C + 0.5 k_1 M_C^2 - g D_C, \\ M_C' = -k_1 M_C^2 - k_2(D_A)M_C + 2g D_C, \\ M_A' = k_2(D_A)M_C - k_3(D_A)M_A - k_4 A M_A - k_5(C_A)M_A^2 + 2g D_A + g C_A, \\ M_N' = k_3(D_A)M_A - d_1 M_N, \\ A' = -k_4 M_A A + g C_A, \\ C_A' = k_4 M_A A - g C_A, \\ D_A' = 0.5 k_5(C_A)M_A^2 - g D_A. \end{cases}$$

For all the simulations, the parameters used, in order to obtain a qualitative result that appears biologically relevant, are provided in Table 1, unless specified otherwise. They are not derived from literature or experimental data but remain in a real range of values. As for initial conditions, they are given in Table 2.

3.1.1 Successive Stresses

Daily stress can be modelled by different successive stresses.

The quantities of monomers M_C and M_A remain slightly above zero because they do not have enough time to fully migrate or reform the crown after stress (see Fig. 6a). As a result, one would observe slight oscillations when zooming in.

We observe a delay before the formation of the second layer of the perinuclear crown (D_A). It only begins to build once the complexes of the first layer are formed, and its development becomes slower as we approach 600 (see Fig. 6b).

The crown reaches the threshold $D_A = 500$ (meaning more than 70% of its total thickness) after approximately 7000 hours.

3.1.2 Continuous Stress

As mentioned above, daily oxidative stress can also be modelled by a continuous and constant stress c_s.

The simulations are very similar to the preceding case. There is always a very small amount of ATM monomers that migrate to the nucleus and we still observe a delay before the formation of the second layer of the perinuclear crown. Nevertheless, the formation of the crown is faster because the stress is considered

Table 1 Description of the parameters and their values used in the model of a cell affected by Alzheimer's disease

Parameters	Descriptions	Values
λ	Constant source of dimers (in h^{-1})	15
d_0	Rate of dimer degradation (in h^{-1})	0.05
d_1	Rate of monomer degradation in the nucleus (in h^{-1})	0.3
k_1	Rate of monomer re-dimerization in the cytoplasm (in h^{-1})	0.01
a_2	Parameter involved in the function k_2 (in h^{-1})	400
b_2	Parameter involved in the function k_2	0.4
n_2	Parameter involved in the function k_2	15
a_3	Parameter involved in the function k_3 (in h^{-1})	80
b_3	Parameter involved in the function k_3	0.5
n_3	Parameter involved in the function k_3	5
k_4	Rate of ATM monomers forming a complex with X proteins (in h^{-1})	0.05
a_5	Parameter involved in the function k_5 (in h^{-1})	0.4
b_5	Parameter involved in the function k_5	150
n_5	Parameter involved in the function k_5	15
c_s	Rate of protein dissociation during constant stress (in h^{-1})	0.002
σ_{sc}	Parameter involved in the Gaussian function of the daily stress	0.3
T	Time between each repetition of stress (in h)	3
a_s	Parameter involved in the Gaussian function of the stress due to irradiation (in h^{-1})	0.8
m_s	Parameter involved in the Gaussian function of the stress due to irradiation	9
σ_s	Parameter involved in the Gaussian function of the stress due to irradiation	2

Table 2 Values for the initial conditions when the crown is not formed

D_{C0}	M_{C0}	M_{N0}	M_{A0}	A_0	C_{A0}	D_{A0}
300	0	0	0	200	0	0

continuous (see Fig. 7b). Indeed, in this case, the crown reaches the threshold $D_A = 500$ after approximately 1700 hours only.

3.2 Influence of the Stress Value on the Formation of the Crown

Not only the way of modelling stress influences the formation of the crown as we have just seen, but also the values of the parameters. Let us focus particularly on the value of c_s when the stress is constant, and observe how the time of formation behaves with respect to this parameter (see Fig. 8). In that way, the time

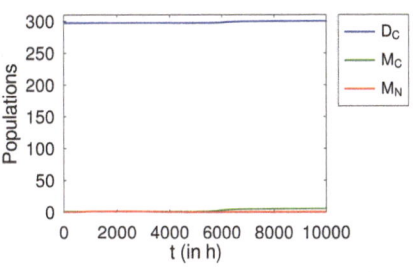 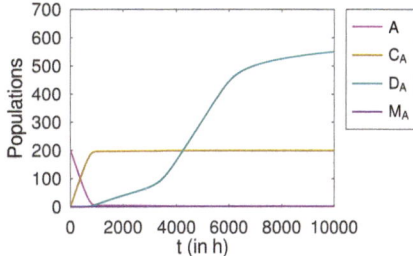

(a) Trajectories of the system. Populations in the cytoplasm and the nucleus.

(b) Trajectories of the system. Populations of the perinuclear crown.

Fig. 6 Simulation of the formation of the perinuclear crown using Gaussian functions. (**a**) Populations in the cytoplasm and the nucleus. (**b**) Populations of the perinuclear crown

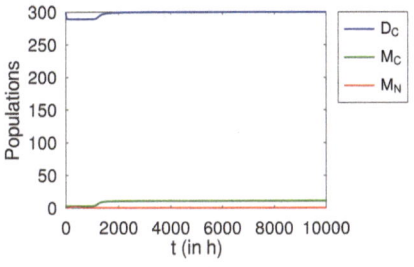 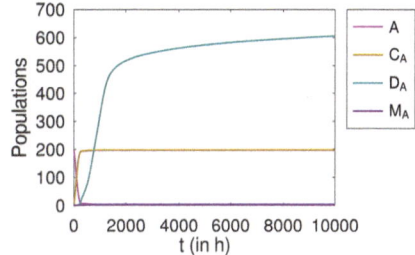

(a) Trajectories of the system. Populations in the cytoplasm and the nucleus.

(b) Trajectories of the system. Populations of the perinuclear crown.

Fig. 7 Simulation of the formation of the perinuclear crown with constant stress. (**a**) Populations in the cytoplasm and the nucleus. (**b**) Populations of the perinuclear crown

Fig. 8 Time for forming the crown t_f (reaching $D_A = 500$) in function of the parameter c_s, with initial conditions of Table 2

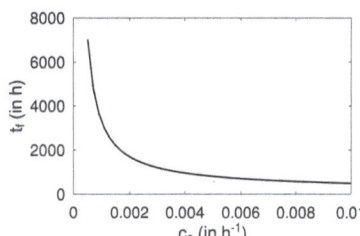

of formation, or more precisely, the time to reach $D_A = 500$, appears to decrease exponentially with the intensity c_s of the stress.

4 Therapeutic Treatment of Alzheimer's Disease

4.1 Impact of Irradiation

4.1.1 Experiments

In the paper in which the perinuclear ATM crowns have been described for the first time [3], the authors investigated the effect of a high dose of ionizing radiation (2 Gy X-rays) on the integrity of the perinuclear ATM crown. Interestingly, a dose of 2 Gy made the great majority of the perinuclear ATM crown disappear in the 10 AD cell lines tested at 10 min post-irradiation. However, progressively, from 4 to 24 h post-irradiation, the number of cells showing perinuclear ATM crowns was found to increase up to the initial value found on the non-irradiated cells. Such observations suggest that irradiation can cause monomerization of the ATM dimers and dissociation of the ATM-APOE complexes. This experimental work was our starting point to induce a therapeutic treatment through irradiation in our model.

4.1.2 Mathematical Model

In order to reduce the perinuclear crown, and dissociate the complexes and the dimers, we simulate irradiation, modelled by a Gaussian function defined for all $t \in \mathbb{R}_+$ by

$$S(t) = a_s e^{-\frac{(t-m_s)^2}{2\sigma_s^2}},$$

with $a_s, m_s, \sigma_s \in \mathbb{R}_+^*$. Here we choose a Gaussian function in order to reflect the progressive effect of the irradiation. We could have also chosen a rectangular pulse, and we would have had the same effect. In order to model the formation of the crown, we keep the constant stress. Thus, the function g is given by $g(S) = c_s + S$ (without irradiation, $S = 0$).

For the simulations with irradiation, we choose the initial conditions in Table 3: the crown is partially formed.

Table 3 Values for the initial conditions when the perinuclear crown is partially formed

D_{C0}	M_{C0}	M_{N0}	M_{A0}	A_0	C_{A0}	D_{A0}
300	0	0	0	0	200	300

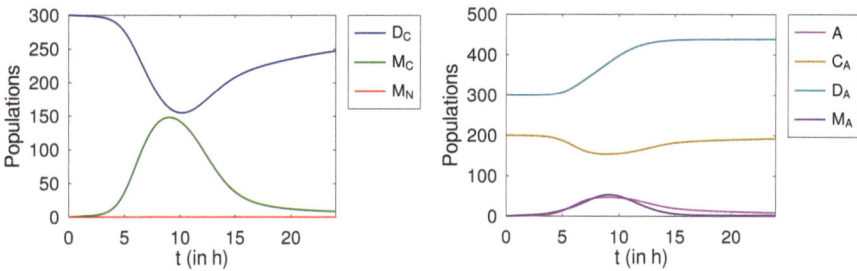

(a) Trajectories of the system. Populations in the cytoplasm and the nucleus.

(b) Trajectories of the system. Populations of the perinuclear crown.

Fig. 9 Simulation of the effect of an irradiation. (**a**) Populations in the cytoplasm and the nucleus. (**b**) Populations of the perinuclear crown

The irradiation causes stronger stress than constant stress, and we observe that indeed, the APOE-ATM complexes dissociate (see Fig. 9). However, these complexes reform immediately, and irradiation also induces the migration of new ATM monomers that reinforce the crown. Nothing crosses the nuclear membrane.

4.2   Impact of Antioxidants

4.2.1   Experiments

In the frame of preliminary experiments, we also investigated the potential effect of some radioprotector drugs on the integrity of the perinuclear ATM crown. However, while the application of some antioxidant agents may influence the number of cells showing perinuclear ATM crowns, the conclusions still need to be consolidated and will be the object of future experimental work.

4.2.2   Mathematical Model

The irradiation is not sufficient to reduce the thickness of the crown. The effect of antioxidants, dispersing the ATM monomers, appears very interesting to prevent the formation of the crown, or even destroy it.

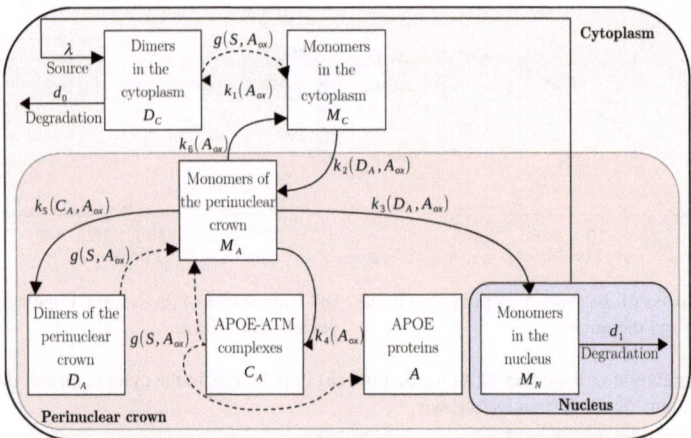

Fig. 10 Diagram of the model for Alzheimer's disease with antioxidants. The functions depend on the stress S due to irradiation and the antioxidants A_{ox}

In this way, we modify the first model (see Fig. 10). The antioxidants act in a way that promotes the passage from compartment M_A to compartment M_C and the passage from compartment M_A to compartment M_N. Denoting A_{ox} the function of the antioxidants, we obtain for all $D_A, A_{ox} \in \mathbb{R}_+$,

$$k_3(D_A, A_{ox}) = b_3 \frac{a_3^{n_3}}{a_3^{n_3} + D_A^{n_3}} (1 + e_3 A_{ox}) \; ; \; k_6(A_{ox}) = e_6 A_{ox},$$

with $a_3, b_3, n_3, e_3, e_6 \in \mathbb{R}_+^*$. On the contrary, they reduce the flow from compartment M_C to compartment M_A and inhibit the formation of complexes. In this way, for all $C_A, D_A, A_{ox} \in \mathbb{R}_+$, we take

$$k_1(A_{ox}) = \frac{a_1}{1 + e_1 A_{ox}} \; ; \; k_2(D_A, A_{ox}) = b_2 \frac{a_2^{n_2}}{(a_2^{n_2} + D_A^{n_2})(1 + e_2 A_{ox})} \; ;$$
$$k_4(A_{ox}) = \frac{a_4}{1 + e_4 A_{ox}} \; ; \; k_5(C_A, A_{ox}) = \frac{a_5 C_A^{n_5}}{(b_5^{n_5} + C_A^{n_5})(1 + e_5 A_{ox})},$$

with $a_1, e_1, a_2, b_2, n_2, e_2, a_4, e_4, a_5, b_5, n_5, e_5 \in \mathbb{R}_+^*$.

Monomerization during oxidative stress is also slowed down as long as the stress is low, which is not the case with irradiation S. We obtain for all $S \in \mathbb{R}_+$ and $A_{ox} \in \mathbb{R}_+$,

$$g(S, A_{ox}) = \frac{c_s}{1 + e_0 A_{ox}} + S,$$

with $c_s, e_0 \in \mathbb{R}_+^*$.
By highlighting the differences in blue, the system becomes:

$$\begin{cases}
D_C' = \lambda - d_0 D_C + 0.5 k_1(A_{ox}) M_C^2 - g(S, A_{ox}) D_C, \\
M_C' = -k_1(A_{ox}) M_C^2 - k_2(D_A, A_{ox}) M_C + k_6(A_{ox}) M_A + 2g(S, A_{ox}) D_C, \\
M_A' = k_2(D_A, A_{ox}) M_C - k_3(D_A, A_{ox}) M_A - k_4(A_{ox}) A M_A - k_5(C_A, A_{ox}) M_A^2 \\
\qquad - k_6(A_{ox}) M_A + 2g(S, A_{ox}) D_A + g(S, A_{ox}) C_A, \\
M_N' = k_3(D_A, A_{ox}) M_A - d_1 M_N, \\
A' = -k_4(A_{ox}) M_A A + g(S, A_{ox}) C_A, \\
C_A' = k_4(A_{ox}) M_A A - g(S, A_{ox}) C_A, \\
D_A' = 0.5 k_5(C_A, A_{ox}) M_A^2 - g(S, A_{ox}) D_A.
\end{cases}$$

The parameters for the simulations are the same as in Table 1 and the new ones are in Table 4.

Table 4 description of the new parameters and their values used in the model of a cell affected by Alzheimer's disease with antioxidants

Parameters	Descriptions	Values
a_1	Rate of monomer re-dimerization in the cytoplasm (in h^{-1})	0.01
a_4	Rate of ATM monomers forming a complex with X proteins (in h^{-1})	0.05
e_0, e_1, e_2, e_4, e_5	Parameters of the antioxidant effect contributing to slow down the process	20
e_3, e_6	Parameters of the antioxidant effect contributing to promoting the process	0.5
c_{ox}	Parameter involved in the Gaussian function of the antioxidant effect (in h^{-1})	1
m_{ox}	Parameter involved in the Gaussian function of the antioxidant effect	9
σ_{ox}	Parameter involved in the Gaussian function of the antioxidant effect	3

Antioxidants as a Means of Preventing the Formation of Crown

We explore the effect of antioxidants daily intake on the cell in order to consider a constant effect, that is to say, for all $t \in \mathbb{R}_+$, $A_{ox}(t) = c_{ox}$.

Taking antioxidants while the crown is not formed (initial conditions of Table 2) helps to prevent its formation. Indeed, the number of ATM dimers of the perinuclear crown remains very low (see Fig. 11).

However, if the antioxidants are taken while the crown is already formed (initial conditions of Table 3), we observe that the thickness (D_A) of the crown decreases, but very slowly (see Fig. 12). Consequently, the cell will die before it will be sufficient to allow a certain quantity of ATM monomers to diffuse in the nucleus.

In this way, when the crown is already constituted, the effect of irradiation alone is not sufficient (the crown reforms immediately) and the effect of antioxidants alone is not enough either. This is why a combination of irradiation and antioxidants is interesting.

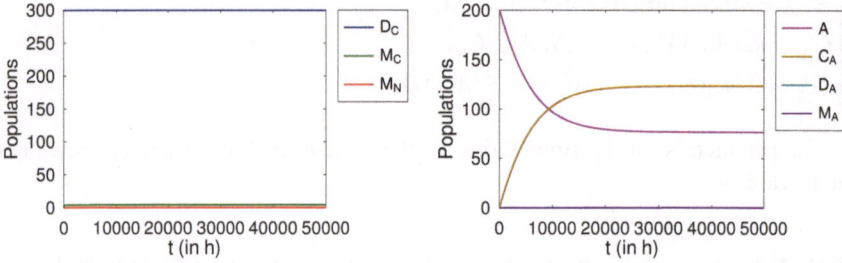

(a) Trajectories of the system. Populations in the cytoplasm and the nucleus.

(b) Trajectories of the system. Populations of the perinuclear crown.

Fig. 11 Simulation with antioxidants on the long term using initial conditions of Table 2. (**a**) Populations in the cytoplasm and the nucleus. (**b**) Populations of the perinuclear crown

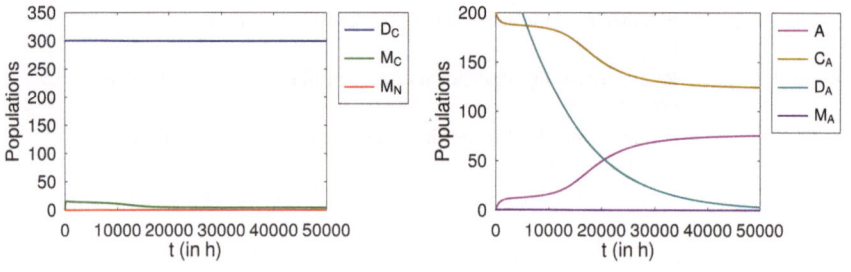

(a) Trajectories of the system. Populations in the cytoplasm and the nucleus.

(b) Trajectories of the system. Populations of the perinuclear crown.

Fig. 12 Simulation with antioxidants on the long term using initial conditions of Table 3. (**a**) Populations in the cytoplasm and the nucleus. (**b**) Populations of the perinuclear crown

4.3 Impact of a Combination of Irradiation and Antioxidants

While with irradiation alone, the perinuclear ATM crown reforms immediately, the addition of antioxidants helps to break the perinuclear ATM crown in the long term by scattering the ATM monomers.

In the Short Term

The aim of antioxidant intake after the irradiation has occurred is precisely to disperse the ATM proteins. First, we simulate a single use of an antioxidant drug, modelled with a Gaussian function defined for all $t \in \mathbb{R}_+$ by

$$A_{ox}(t) = c_{ox}e^{-\frac{(t-m_{ox})^2}{2\sigma_{ox}^2}},$$

with $c_{ox}, m_{ox}, \sigma_{ox} \in \mathbb{R}_+^*$.

We suppose the crown to be partially formed (initial conditions of Table 3). After simulation, we observe firstly that thanks to the antioxidants, a large number of monomers in the cytoplasm re-dimerize, and thus less monomers migrate (see Fig. 13a). Secondly, antioxidants help break down the crown, primarily the layer of ATM dimers (see Fig. 13b), allowing a certain number of monomers to pass into the nucleus.

In the Long Term

Reformation of Perinuclear Crown

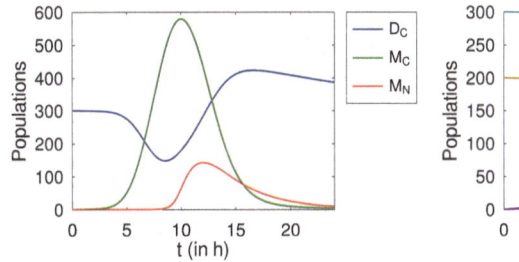 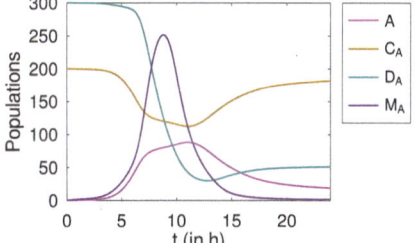

(a) Trajectories of the system. Populations in the cytoplasm and the nucleus.

(b) Trajectories of the system. Populations of the perinuclear crown.

Fig. 13 Simulation with irradiation and antioxidants in the short term, when the crown is partially formed. **(a)** Populations in the cytoplasm and the nucleus. **(b)** Populations of the perinuclear crown

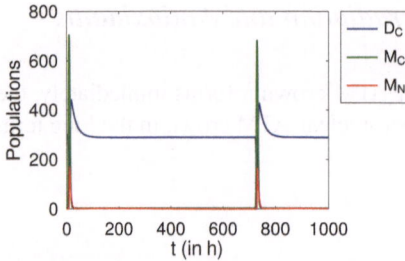 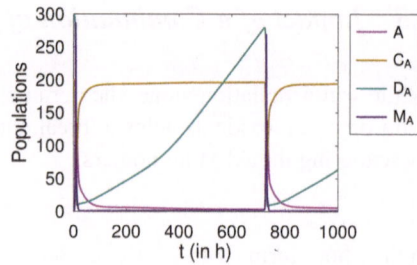

(a) Trajectories of the system. Populations in the cytoplasm and the nucleus.

(b) Trajectories of the system. Populations of the perinuclear crown.

Fig. 14 Simulation with the effect of irradiation and antioxidants repeated in the long term. (**a**) Populations in the cytoplasm and the nucleus. (**b**) Populations of the perinuclear crown

Fig. 15 Time for reformation of the crown t_{ref} (reaching $D_A = 350$) in function of the dose of antioxidants A_{ox}, with initial conditions of Table 3

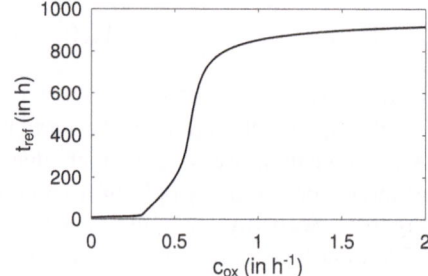

On the longer term, we observe the regeneration of the perinuclear crowns. We then repeat the irradiation combined with antioxidants, which once again reduces the perinuclear ATM crowns (see Fig. 14) and may eventually save the cell from a fatal issue. This, considered on a larger scale and applied to a whole tissue zone (like the hippocampus) may have a positive therapeutic effect on the disease. The patient, if not totally cured, may live longer without memory impairment.

Let us focus on the necessary time to reform (partially) the perinuclear ATM crown after irradiation and using antioxidants. To this aim, we study the time to reach $D_A = 350$ for different doses of antioxidants represented by c_{ox} (see Fig. 15). We realize that antioxidants have very low effectiveness at the time of reformation when the dose is below approximately 0.3: the perinuclear ATM crown reforms immediately. Moreover, once we reach around 0.8, we achieve an optimal effect, that is to say, that the perinuclear ATM crown reforms slowly and increasing the dose will have a minimal impact on the time of reformation.

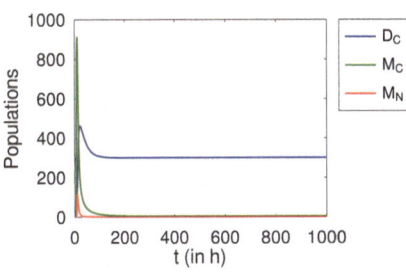
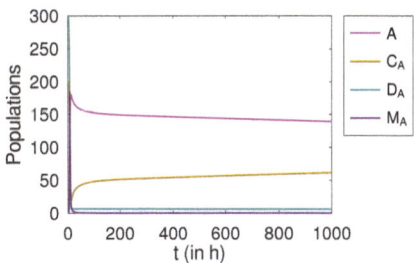

(a) Trajectories of the system. Populations in the cytoplasm and the nucleus.

(b) Trajectories of the system. Populations of the perinuclear crown.

Fig. 16 Simulation with irradiation and antioxidants where the effect of the antioxidant is prolonged. (**a**) Populations in the cytoplasm and the nucleus. (**b**) Populations of the perinuclear crown

Prolongation of the Effect of the Antioxidants

It is also possible, instead of repeating the irradiation, to prolong the effect of antioxidants so that the perinuclear ATM crown does not reconstitute, as in Sect. "Antioxidants as a Means of Preventing the Formation of Crown", that is to say, we consider the effect to be constant: $A_{ox}(t) = c_{ox}$ for all $t \in \mathbb{R}_+$. In that case, the amount of ATM dimers of the perinuclear crown remains low (see Fig. 16). A quantity of monomers diffuses into the nucleus during the irradiation and then decreases and remains low due to the low oxidative stress.

5   Discussion and Conclusion

Utilizing the same mathematical model, we have delineated the formation of the perinuclear ATM crown within cells affected by Alzheimer's disease and explored the noteworthy impact of irradiations and antioxidants on these cells.

Nevertheless, our model stands amenable to refinement. Initially, it postulates a finite growth limit for the perinuclear crown. However, in vitro, the termination of the cell's viability precedes the comprehensive observation of perinuclear crown formation behaviour. Additionally, throughout the simulations, an assumption was made regarding the identical stress response for both ATM dimers and ATM-APOE complexes. Moreover, simulations were exclusively conducted for specific coefficient values denoted as e_i, where i ranges from 0 to 6. To elucidate the influence of these parameters, especially in the context of prolonged antioxidant application (see Fig. 17) and the combined application of irradiation and antioxidants (see Fig. 18), a more comprehensive exploration is warranted.

For several sets of parameters, we plot the trajectories for different initial conditions by varying D_{A0} (Table 5).

We take the parameter set 1 as the reference (see Fig. 17a), which was used in the previous sections, and we vary:

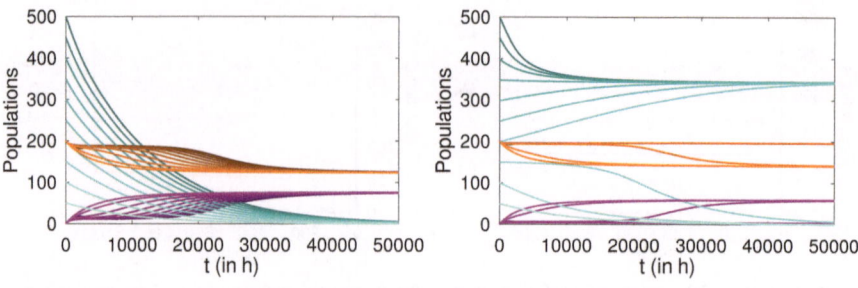

(a) Parameter set 1: $e_0 = 20$, $e_1 = 20$, $e_2 = 20$, $e_3 = 0.5$, $e_4 = 20$, $e_5 = 20$, $e_6 = 0.5$, $c_{ox} = 1$.

(b) Parameter set 2: $e_0 = 20$, $e_1 = 20$, $e_2 = 20$, $e_3 = 0.1$, $e_4 = 20$, $e_5 = 20$, $e_6 = 0.1$, $c_{ox} = 1$.

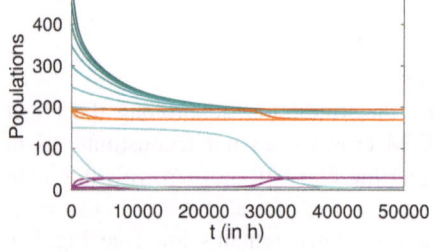
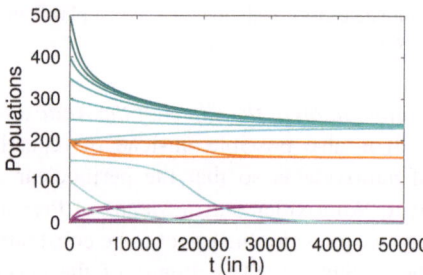

(c) Parameter set 3: $e_0 = 5$, $e_1 = 5$, $e_2 = 5$, $e_3 = 0.5$, $e_4 = 5$, $e_5 = 5$, $e_6 = 0.5$, $c_{ox} = 1$.

(d) Parameter set 4: $e_0 = 20$, $e_1 = 20$, $e_2 = 20$, $e_3 = 0.5$, $e_4 = 20$, $e_5 = 20$, $e_6 = 0.5$, $c_{ox} = 0.5$.

Fig. 17 Simulations of the effect of antioxidants $A_{ox}(t) = c_{ox}$, $t > 0$, for multiple parameter sets (**a**), (**b**), (**c**) and (**d**), and for each set, different initial conditions for D_A. In magenta, population A; in orange, population C_A; in blue, population D_A

Fig. 18 Simulation with irradiation and antioxidants where the effect of the antioxidant is prolonged, using the set of parameters 4

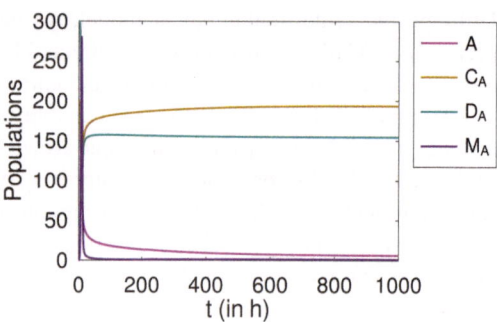

Table 5 Values of the initial conditions for Fig. 17

D_{C0}	M_{C0}	M_{N0}	M_{A0}	A_0	C_{A0}	D_{A0}
300	0	0	0	0	200	from 0 to 500, incrementing by 50

- the parameters contributing to slowing down the process $(e_0, e_1, e_2, e_4, e_5)$ by reducing them from 20 to 5 (see Fig. 17b),
- the parameters contributing to accelerate the process (e_3, e_6) by reducing them from 0.5 to 0.1 (see Fig. 17c),
- the parameter c_{ox}, by reducing it from 1 to 0.5, which is equivalent to divide by 2 the parameters e_i, $i = 1, \ldots, 6$ (see Fig. 17d).

According to the parameters and therefore the effect of antioxidants, there is *a priori* one or two asymptotically stable equilibria (for the other initial conditions fixed). When the effect of antioxidants is weaker, beyond a certain threshold value of D_{A0}, the perinuclear ATM crown is not able to fully disappear (see Fig. 17b, 17c and 17d).

Even with irradiation followed by antioxidants like in Sect. "In the Long Term", it is insufficient to eliminate the perinuclear ATM crown (see Fig. 18).

In conclusion, the model presented herein constitutes an initial qualitative approximation. The conjecture that the perinuclear ATM crown serves as a distinctive marker for Alzheimer's disease holds the potential to establish criteria for identifying a potential predisposition to the disease. Subsequent validation and quantitative exploration of the model will be feasible upon the availability of patient data and experimental results. Comparative analysis with experimental findings will facilitate an in-depth comprehension of the disease's progression and treatment possibilities, particularly concerning dosage parameters such as antioxidant doses and administration frequencies. This avenue constitutes the focus of our forthcoming investigations.

Consequently, the overarching objective encompasses three facets: firstly, to furnish an early diagnosis facilitating the prevention of crown formation through antioxidant application; secondly, to diagnose and manage the disease in instances where crowns have already formed; and thirdly, to efficaciously predict the progression of the crown, particularly during therapeutic interventions. Furthermore, the adaptability of this model to the specificities of other diseases validated under the Radiation-Induced NucleoShuttling of the ATM protein (RIANS) framework, such as Huntington's disease, tuberous sclerosis complex syndrome, Xeroderma Pigmentosum, neurofibromatosis type 1, and Rothmund–Thomson syndrome, is envisaged [1, 7–10].

Competing Interests The authors have no conflicts of interest to declare that are relevant to the content of this chapter.

References

1. Al-Choboq, J., Nehal, M., Sonzogni, L., Granzotto, A., El Nachef, L., Restier-Verlet, J., Maalouf, M., Berthel, E., Aral, B., Corradini, N., Bourguignon, M., Foray, N.: Molecular and cellular responses to ionization radiation in untransformed fibroblasts from the Rothmund–Thomson syndrome: influence of the nucleo-shuttling of the ATM protein kinase. Radiation **3**, 21–38 (2023). https://doi.org/10.3390/radiation3010002

2. Berthel, E., Foray, N., Ferlazzo, M.: The nucleoshuttling of the ATM protein: a unified model to describe the individual response to high- and low-dose of radiation? Cancers **11**, 905 (2019). https://doi.org/10.3390/cancers11070905

3. Berthel, E., Pujo-Menjouet, L., Le Reun, E., Sonzogni, L., Al-Choboq, J., Chekroun, A., Granzotto, A., Duclos, M., Devic, C., Ferlazzo, M.L., Pereira, S., Bourguignon, M., Foray, N.: Toward an early diagnosis for Alzheimer's disease based on the perinuclear localization of the ATM protein. Cells **12**(13), 1747 (2023). https://doi.org/10.3390/cells12131747

4. Bodgi, L., Foray, N.: The nucleo-shuttling of the ATM protein as a basis for a novel theory of radiation response: resolution of the linear-quadratic model. Int. J. Radiat. Biol. **92**, 117–131 (2016). https://doi.org/10.3109/09553002.2016.1135260

5. Bodgi, L., Canet, A., Pujo-Menjouet, L., Lesne, A., Victor, J.M., Foray, N.: Mathematical models of radiation action on living cells: from the target theory to the modern approaches. A historical and critical review. J. Theor. Biol. **394**, 93–101 (2016). https://doi.org/10.1016/j.jtbi.2016.01.018

6. Breijyeh, Z., Karaman, R.: Comprehensive review on Alzheimer's disease: causes and treatment. Molecules **25**(24), 5789 (2020). https://doi.org/10.3390/molecules25245789

7. Combemale, P., Sonzogni, L., Devic, C., Bencokova, Z., Ferlazzo, M., Granzotto, A., Burlet, S., Pinson, S., Amini-Adle, M., Al-Choboq, J., Bodgi, L., Bourguignon, M., Balosso, J., Bachelet, J., Foray, N.: Individual response to radiation of individuals with neurofibromatosis type I: role of the ATM protein and influence of statins and bisphosphonates. Mol. Neurobiol. **59**, 556–573 (2022). https://doi.org/10.1007/s12035-021-02615-3

8. Ferlazzo, M., Sonzogni, L., Granzotto, A., Bodgi, L., Lartin, O., Devic, C., Vogin, G., Pereira, S., Foray, N.: Mutations of the Huntington's disease protein impact on the ATM-dependent signaling and repair pathways of the radiation-induced DNA double-strand breaks: corrective effect of statins and bisphosphonates. Mol. Neurobiol. **49**, 1200–1211 (2014). https://doi.org/10.1007/s12035-013-8591-7

9. Ferlazzo, M., Bach-Tobdji, M.K.E., Djerad, A., Sonzogni, L., Devic, C., Granzotto, A., Bodgi, L., Bachelet, J., Djefal-Kerrar, A., Hennequin, C., Foray, N.: Radiobiological characterization of tuberous sclerosis: a delay in the nucleo-shuttling of ATM may be responsible for radiosensitivity. Mol. Neurobiol. **55**, 4973–4983 (2018). https://doi.org/10.1007/s12035-017-0648-6

10. Ferlazzo, M., Berthel, E., Granzotto, A., Devic, C., Sonzogni, L., Bachelet, J.T., Pereira, S., Bourguignon, M., Sarasin, A., Mezzina, M., Foray, N.: Some mutations in the xeroderma pigmentosum D gene may lead to moderate but significant radiosensitivity associated with a delayed radiation-induced ATM nuclear localization. Int. J. Radiat. Biol. **96**, 394–410 (2020). https://doi.org/10.1080/09553002.2020.1694189

11. Ferrell, J.: Tripping the switch fantastic: how protein kinase cascade convert graded into switch-like outputs. Trends Biochem. Sci. **21**, 460–466 (1996). https://doi.org/10.1016/s0968-0004(96)20026-x

12. Kellerer, A., Rossi, H.: Theory of dual radiation action. Curr. Top. Radiat. Res. Quart. **8**(2), 85–158 (1972). https://www.osti.gov/biblio/4611340

13. Kellerer, A., Rossi, H.: A generalized formulation of dual radiation action. Radiat. Res. **75**(3), 471–488 (1978). https://doi.org/10.2307/3574835

14. Le Reun, E., Bodgi, L., Granzotto, A., Sonzogni, L., Ferlazzo, M.L., Al-Choboq, J., El-Nachef, L., Restier-Verlet, J., Berthel, E., Devic, C., Bouchet, A., Bourguignon, M., Foray, N.: Quantitative correlations between radiosensitivity biomarkers show that the ATM protein kinase is strongly involved in the radiotoxicities observed after radiotherapy. Int. J. Mol. Sci. **23**(18), 10434 (2022). https://doi.org/10.3390/ijms231810434
15. Lloret, A., Esteve, D., Lloret, M.A., Cervera-Ferri, A., Lopez, B., Nepomuceno, M., Monllor, P.: When does alzheimer's disease really start? The role of biomarkers. Int. J. Mol. Sci. **20**(22), 941–956 (2019). https://doi.org/10.3390/ijms20225536
16. Mackey, M.: Unified hypothesis of the origin of aplastic anemia and periodic hematopoiesis. Blood **51**, 941–956 (1978). https://doi.org/10.1182/blood.V51.5.941.941
17. Šerý, O., Povová, J., Míšek, I., Pešák, L., Janout, V.: Review paper Molecular mechanisms of neuropathological changes in Alzheimer's disease: a review. Folia Neuropathol. **51**(1), 1–9 (2013). https://doi.org/10.5114/fn.2013.34190
18. World Health Organization: Fact sheets of dementia (2023). https://www.who.int/news-room/fact-sheets/detail/dementia

Ergodic and Chaotic Properties of Some Biological Models

Ryszard Rudnicki ⓘ

Abstract In this note we present two types of biological models which have interesting ergodic and chaotic properties. The first type are one-dimensional transformations, like a logistic map, which are used to describe the change in population size in successive generations. We study ergodic properties of such transformations using Frobenius–Perron operators. The second type are some structured populations models, for example a space-structured model, or a model of maturity-distribution of precursors of blood cells. These models are described by partial differential equations, which generate semiflows on the space of functions. We construct strong mixing invariant measures for these semiflows using stochastic precesses. From properties of invariant measures we deduce some chaotic properties of semiflows such as the existence of dense trajectories and strong instability of all trajectories.

1 Introduction

Many biological processes behave in a chaotic manner. Classic examples are the discrete time population growth model $x_{n+1} = 4x_n(1 - x_n)$; the Mackey–Glass model [1], in which the number of red blood cells changes according to a delay differential equation; and a maturity structured model [2], in which a partial differential equation describes the evolution of the distribution of cell maturity. One method of studying chaos is to use the ergodic theory. In general, the stronger the ergodic properties of a dynamical system, the more chaotic its trajectories are (see Sect. 4).

The aim of this note is to present some results concerning ergodic and chaotic properties of two types of biological models. The first type are discrete time population growth models. Such models are described by unimodal transformations and a very useful tool for studying their ergodic properties are the Frobenius–Perron

R. Rudnicki (✉)
Institute of Mathematics, Polish Academy of Sciences, Katowice, Poland
e-mail: ryszard.rudnicki@us.edu.pl

© The Author(s), under exclusive license to Springer Nature Switzerland AG 2025
Y. Mori et al. (eds.), *Dynamics of Physiological Control*, Lecture Notes
on Mathematical Modelling in the Life Sciences,
https://doi.org/10.1007/978-3-031-82396-1_6

operators [3]. In the next section, we briefly recall the definitions of the basic ergodic properties and their relation to Frobenius–Perron operators. In Sect. 3 we formulate and prove a general result concerning the existence of invariant measures for a large class of unimodal transformations (see Theorem 4). The invariant measure is absolutely continuous with respect to the Lebesgue measure, its density is a positive and continuous function and the transformation is exact. We apply this theorem to the transformation $S(x) = cx(1 - x^2)$, to the Beverton–Holt transformation $S(x) = -ax + \frac{bx}{1+x}$, and to the Ricker transformation $S(x) = -ax + axe^{\lambda(K-x)}$.

Section 4 is devoted to study ergodic and chaotic properties of some models of structured populations. Such models are usually described by partial differential equations, and thus the dynamical systems generated by these equations are defined on infinite-dimensional spaces. When looking for an invariant measure for an infinite dimensional dynamical system, we do not use Frobenius–Perron operators, because it is difficult to indicate the measure with respect to which the invariant measure would be absolutely continuous (would have a density). If we consider partial differential equations in which the spatial variable x is a real number, it has turned out convenient to construct an invariant measure by means of stationary stochastic processes and isomorphism of the initial dynamic system with the shift $(T^t \varphi)(s) = \varphi(s + t)$ on a properly chosen space. A major challenge is to construct an invariant measure if x is multidimensional. Then instead of a stationary process, we use a random field [4]. We show how to prove that such dynamical systems have exact invariant measures supported on the whole space. We also present conclusions on the chaotic behavior of such systems. We apply the obtained results to various biological models: maturity-distribution of precursors of blood cells; a size structured cellular population; and the space dispersal of a population.

2 Measure-Preserving Dynamical Systems

Let (X, Σ, μ) be a probability space, i.e. μ is a probability measure defined on a σ-algebra Σ of subsets of X. Let $\mathcal{T} = \mathbb{N} = \{0, 1, \dots\}$ or $\mathcal{T} = [0, \infty)$. *One-parameter semigroup* is a family $\{\pi_t\}_{t \in \mathcal{T}}$ of transformations of X such that $\pi_s \circ \pi_t = \pi_{s+t}$ for $s, t \in \mathcal{T}$ and $\pi_0 = \text{Id}$. If additionally the function $(t, x) \mapsto \pi_t x$ is measurable as a function from the Cartesian product $\mathcal{T} \times X$ to X and the measure μ is *invariant* with respect to each transformation $\pi_t, t \in \mathcal{T}$, i.e. $\mu(\pi_t^{-1}(A)) = \mu(A)$ for any set $A \in \Sigma$, then the quadruple (X, Σ, μ, π_t) is called a *measure-preserving dynamical system*. The triple (X, Σ, μ) is called a *phase space* and the set $\{\pi_t x : t \in \mathcal{T}\}$ is called the *trajectory of a point* $x \in X$. If $\mathcal{T} = \mathbb{N}$, $S: X \to X$ is a measurable transformation, the measure μ is invariant with respect to S, and $\pi_t = S^t$ is the t-th iterate of S, then (X, Σ, μ, π_t) is a measure-preserving dynamical system.

Our goal is to study properties of measure-preserving dynamical systems: ergodicity, mixing and exactness. A set A is called *invariant* with respect to a semigroup $\{\pi_t\}_{t \in \mathcal{T}}$, if $\pi_t^{-1}(A) = A$ for any $A \in \Sigma$ and $t \in \mathcal{T}$. The family of invariant sets forms a σ-algebra Σ_{inv}. If σ-algebra Σ_{inv} is trivial, i.e. it consists only of sets of measure μ zero or one, then the measure-preserving dynamical system is said to be *ergodic*.

A stronger property than ergodicity is mixing. A measure-preserving dynamical system (X, Σ, μ, π_t) is called *mixing* if

$$\lim_{t \to \infty} \mu(A \cap \pi_t^{-1}(B)) = \mu(A)\mu(B) \quad \text{for all } A, B \in \Sigma. \tag{1}$$

Identifying the measure μ with the probability P, one can formulate the condition (1) as follows:

$$\lim_{t \to \infty} P(\pi_t(x) \in A | x \in B) = P(A) \quad \text{for all } A, B \in \Sigma \text{ and } P(B) > 0,$$

which means that the trajectories of almost all points enter a set A with asymptotic probability $P(A)$.

A stronger property than mixing is exactness. A system (X, Σ, μ, π_t) with double measurable transformations π_t, i.e. $\pi_t(A) \in \Sigma$ and $\pi_t^{-1}(A) \in \Sigma$ for all $A \in \Sigma$ and $t \in \mathcal{T}$, and with an invariant probability measure μ is called *exact* if for every set $A \in \Sigma$ with $\mu(A) > 0$ we have $\lim_{t \to \infty} \mu(\pi_t(A)) = 1$. Exactness is equivalent to the following condition: σ-algebra $\bigcap_{t \geq 0} \pi_t^{-1}(\Sigma)$ contains only sets of measure μ zero or one. Here $\pi_t^{-1}(\Sigma) = \{\pi_t^{-1}(A) : A \in \Sigma\}$.

The ergodic properties of the dynamical system can be studied using Frobenius–Perron operators [3]. Let (X, Σ, m) be a σ-finite measure space. A measurable map $\varphi: X \to X$ is called *non-singular* if it satisfies the following condition

$$m(A) = 0 \implies m(\varphi^{-1}(A)) = 0 \text{ for } A \in \Sigma. \tag{2}$$

Let $L^1 = L^1(X, \Sigma, m)$ and let φ be a measurable nonsingular transformation of X. An operator $P_\varphi: L^1 \to L^1$ which satisfies the following condition

$$\int_A P_\varphi f(x) \, m(dx) = \int_{\varphi^{-1}(A)} f(x) \, m(dx) \text{ for } A \in \Sigma \text{ and } f \in L^1 \tag{3}$$

is called the *Frobenius–Perron operator* for the transformation φ. The operator P_φ is linear, *positive* (if $f \geq 0$ then $P_\varphi f \geq 0$) and preserves the integral ($\int_X P_\varphi f \, dm = \int_X f \, dm$). The adjoint of the Frobenius–Perron operator $P_\varphi^*: L^\infty \to L^\infty$ is given by $P_\varphi^* g(x) = g(\varphi(x))$.

Let $\varphi: X \to X$ be a nonsingular transformation of a σ-finite measure space (X, Σ, m). Denote by D the set of all densities with respect to m, i.e. functions $f \in L^1(X, \Sigma, m)$ such that $f \geq 0$ and $\|f\| = 1$. Let $f^* \in D$. Then the measure $\mu(A) = \int_A f^* \, dm$ for $A \in \Sigma$, is invariant with respect to φ if and only if $P_\varphi f^* = f^*$. If the map $\pi: \mathcal{T} \times X \to X$ is measurable, $\{\pi_t\}_{t \in \mathcal{T}}$ is a one-parameter semigroup of nonsingular transformations of (X, Σ, m), and P^t denotes the Frobenius–Perron operator corresponding to π_t, then the quadruple (X, Σ, μ, π_t) is a measure-preserving dynamical system if and only if $P^t f^* = f^*$

Table 1 The relations between ergodic properties of dynamical systems (X, Σ, μ, π_t) and the behavior of the Frobenius–Perron operators P^t

μ	f^*
Invariant	$P^t f^* = f^*$ for all $t \in \mathcal{T}$
Ergodic	f^* is a unique fixed point in D of all P^t
Mixing	w-$\lim_{t \to \infty} P^t f = f^*$ for every $f \in D$
Exact	$\lim_{t \to \infty} P^t f = f^*$ for every $f \in D$

for all $t \in \mathcal{T}$. We collect the relations between ergodic properties of dynamical systems (X, Σ, μ, π_t) and the behavior of the Frobenius–Perron operators P^t in Table 1.

We recall that the *weak limit* w-$\lim_{t \to \infty} P^t f$ is a function $h \in L^1$ such that for every $g \in L^\infty$ we have

$$\lim_{t \to \infty} \int_X P^t f(x) g(x) \, m(dx) = \int_X h(x) g(x) \, m(dx).$$

Now we show how to find the Frobenius–Perron operator for a piecewise smooth transformation φ of some interval Δ. We assume that there exists at most countable family of pairwise disjoint open intervals Δ_i, $i \in I$, contained in Δ and having the following properties:

(a) the sets $\Delta_0 = \Delta \setminus \bigcup_{i \in I} \Delta_i$ and $\varphi(\Delta_0)$ have zero Lebesgue measure,
(b) maps $\varphi_i = \varphi\big|_{\Delta_i}$ are C^1-maps from Δ_i onto $\varphi(\Delta_i)$ and $\varphi_i'(x) \neq 0$ for $x \in \Delta_i$.

Then the Frobenius–Perron operator P_φ exists and is given by the formula

$$P_\varphi f(x) = \sum_{i \in I_x} f(\psi_i(x)) |\psi_i'(x)|, \tag{4}$$

where $\psi_i = \varphi_i^{-1}$ and $I_x = \{i : x \in \varphi(\Delta_i)\}$ (see [5, 6]).

Example 1 Let us consider the *tent map* $\varphi : [0, 1] \to [0, 1]$ given by

$$\varphi(x) = \begin{cases} 2x & \text{for } x \in [0, 1/2], \\ 2 - 2x & \text{for } x \in (1/2, 1]. \end{cases} \tag{5}$$

The Frobenius–Perron operator P_φ is of the form

$$P_\varphi f(x) = \tfrac{1}{2} f(\tfrac{1}{2}x) + \tfrac{1}{2} f(1 - \tfrac{1}{2}x). \tag{6}$$

It is easy to see that the density $f^* = 1_{[0,1]}$ satisfies $P_\varphi f^* = f^*$, and therefore the Lebesgue measure on $[0, 1]$ is invariant with respect φ. We check that $\lim_{t \to \infty} P_\varphi^t f = f^*$ for any density f. It is sufficient to check this condition for densities which are Lipschitz continuous. Let L be the Lipschitz constant for f. Then

$$|P_\varphi f(x) - P_\varphi f(y)| \leq \tfrac{1}{2}|f(\tfrac{x}{2}) - f(\tfrac{y}{2})| + \tfrac{1}{2}|f(1 - \tfrac{1}{2}x) + f(1 - \tfrac{1}{2}y)| \leq \tfrac{L}{2}|x - y|.$$

Thus $L/2$ is the Lipschitz constant for $P_\varphi f$ and by induction we conclude that $L/2^t$ is the Lipschitz constant for $P_\varphi^t f$. Hence, the sequence $(P_\varphi^t f)$ converges uniformly to a constant function. Since $P_\varphi^t f$ are densities, $(P_\varphi^t f)$ converges to f^* uniformly, which implies the convergence in L^1. The condition $\lim_{t\to\infty} P_\varphi^t f = f^*$ implies the exactness of the transformation φ.

We now introduce the notion of an isomorphic dynamical system, which we use extensively in the next section.

Let (X, Σ, μ, π_t) be a measure-preserving dynamical system, $(\tilde{X}, \tilde{\Sigma})$ be a measurable space and $\tilde{\pi}: \mathcal{T} \times \tilde{X} \to \tilde{X}$ be a measurable one-parameter semigroup. Assume that there exists an invertible double measurable function $\alpha: X \to \tilde{X}$ such that $\alpha \circ \pi_t = \tilde{\pi}_t \circ \alpha$ for $t \in \mathcal{T}$. Let $\tilde{\mu}(A) = \mu(\alpha^{-1}(A))$ for $A \in \tilde{\Sigma}$. Then $(\tilde{X}, \tilde{\Sigma}, \tilde{\mu}, \tilde{\pi}_t)$ is a measure-preserving dynamical system *isomorphic* to (X, Σ, μ, π_t). If two dynamical systems are isomorphic, then they have the same ergodic properties.

Example 2 Now we check that the *logistic map* $\psi(x) = 4x(1 - x)$ on $[0, 1]$ is exact. Let φ be the tent map and $\alpha(x) = \tfrac{1}{2} - \tfrac{1}{2}\cos(\pi x)$. Then $\psi \circ \alpha = \alpha \circ \varphi$. Let $\mu(A) = |\alpha^{-1}(A)|$ for $A \in \mathcal{B}([0, 1])$, where $|\cdot|$ is the Lebesgue measure on the σ-algebra $\mathcal{B}([0, 1])$ of Borel subsets of $[0, 1]$. This implies that the map ψ is an exact transformation of the space $([0, 1], \mathcal{B}([0, 1]), \mu)$. Since $\alpha^{-1}(x) = \pi^{-1}\arccos(1 - 2x)$ we find that $d\mu = g^*(x)\,dx$, where

$$g^*(x) = \frac{d}{dx}\alpha^{-1}(x) = \left[\pi\sqrt{x(1 - x)}\right]^{-1}.$$

More advanced examples of applications of Frobenius–Perron operators to study ergodic properties of dynamical systems can be found in [3, 7, 8].

3   Population Generation Models

Consider a population model in which the number of individuals in successive generations changes according to a recursive formula

$$x_{n+1} = S(x_n), \tag{7}$$

$S: [0, K] \to [0, K]$ is a continuous function of the form $S(x) = bxf(x)$, where b is the *per capita birth rate*, K is the *capacity of the environment*, in this case we identify it with the maximum population size, and $f(x)$ is a *competitive function* describing the probability of survival if x is the population size. This is a *model with disjoint generations*. We assume that for some $\tilde{x} \in (0, K]$ we have $S(\tilde{x}) = K$, because otherwise the maximum population size would be smaller than K.

We will consider specific transformations S appearing in known models, in which $S(0) = S(K) = 0$ and S is a strictly increasing function in the interval $[0, \tilde{x}]$ and strictly decreasing in the interval $[\tilde{x}, K]$. Such transformations are called *unimodal*.

Example 2 suggests that exactness of a unimodal transformation $S \colon [0, K] \to [0, K]$ can be studied by showing the isomorphism of S with the tent map φ, and thus finding such a function Φ strictly increasing from the interval $[0, 1]$ to the interval $[0, K]$ (preferably of class C^1) satisfying the condition $S \circ \Phi = \Phi \circ \varphi$. Unfortunately, this method is not easy, so we need to modify it. To do this, we present a theorem on the exactness of a certain class of piecewise expanding maps of an interval $[0, L]$.

Theorem 3 *Let $S \colon [0, L] \to [0, L]$ be a map satisfying the following conditions:*

   (i) *there exists a partition $0 = a_0 < a_1 < \cdots < a_r = L$ of the interval $[0, L]$ such that for each $i = 1, \ldots, r$ the restriction of S to the interval (a_{i-1}, a_i) is a C^2-function,*
  (ii) *$S((a_{i-1}, a_i)) = (0, L)$ for $i = 1, \ldots, r$,*
 (iii) *$|S'(x)| > \lambda$ for some $\lambda > 1$ and for all $x \neq a_i$, $i = 0, \ldots, r$,*
 (iv) *$|S''(x)|/[S'(x)]^2 < c$ for some $c > 0$ and for all $x \neq a_i$, $i = 0, \ldots, r$.*

Then the Frobenius–Perron operator P_S is asymptotically stable, i.e., there exists $f^* \in D$ such that $\lim_{n \to \infty} P_S^n f = f^*$ for every $f \in D$.

Remark 1 If the map S satisfies conditions (i), (iii), and

 (iv') the map S has C^2 extensions from open intervals (a_{i-1}, a_i) onto closed intervals $[a_{i-1}, a_i]$, for $i = 1, \ldots, r$,

then S has an invariant measure absolutely continuous with respect to the Lebesgue measure (see [9]). The condition (iv) follows immediately from (iv'). Moreover if S satisfies condtions (i), (ii), (iii), (iv'), then the invariant density f^* is a continuous function and satisfies inequalities

$$M^{-1} \leq f^*(x) \leq M \quad \text{for some } M > 0 \text{ and for each } x \in [0, L]. \tag{8}$$

The proof of Theorem 3 is given in [3, Th. 6.2.2]. The proof of the property (8), as well as a number of other interesting results on one-dimensional transformations can be found in the monograph [7].

Now we give sufficient conditions for exactness of unimodal transformations.

Theorem 4 *We assume that S is a C^3 unimodal function satisfying the condition*

$$S'(x) > 0 \text{ for } x \in [0, \tilde{x}), \quad S'(x) < 0 \text{ for } x \in (\tilde{x}, K], \quad S''(\tilde{x}) < 0. \tag{9}$$

Let

$$h_1(x) = \sqrt{\frac{x(K-x)}{S(x)}}, \quad h_2(x) = \frac{|S'(x)|}{\sqrt{(K-S(x))}}. \tag{10}$$

If $\inf h_1 h_2 > 1$, *then there is a probability measure* μ *absolutely continuous with respect to the Lebesgue measure such that the quadruple* $([0, K], \mathcal{B}([0, K]), \mu, S)$ *is a measure-preserving dynamical system. This system is exact, and the density* g^* *of the measure* μ *is a positive and continuous function on the interval* $(0, K)$ *and*

$$\lim_{x \to 0^+} g^*(x) = \lim_{x \to K^-} g^*(x) = \infty. \tag{11}$$

Proof Let Φ be the function from the interval $[0, \pi]$ onto the interval $[0, K]$ given by $\Phi(x) = \frac{K}{2}(1 - \cos x)$. Then Φ is strictly increasing and we can define the transformation $\tilde{S} \colon [0, \pi] \to [0, \pi]$ by $\tilde{S}(x) = \Phi^{-1} \circ S \circ \Phi$. We check that $\tilde{S}$ satisfies conditions (i), (ii), (iii), and (iv'). Let $\Psi = \Phi^{-1}$ and $\psi = \Psi'$. Then

$$\tilde{S}'(x) = \Psi'((S \circ \Phi)(x)) S'(\Phi(x)) \Phi'(x) = \frac{\Psi'(S(y)) S'(y)}{\Psi'(y)} = \frac{\psi(S(y)) S'(y)}{\psi(y)},$$

where $y = \Phi(x)$. Let

$$h(y) = \frac{\psi(S(y)) S'(y)}{\psi(y)}.$$

Then

$$\tilde{S}''(x) = h'(\Phi(x)) \Phi'(x) = \frac{h'(y)}{\psi(y)}.$$

Since

$$\Psi(x) = \Phi^{-1}(x) = \arccos(1 - 2x/K), \quad \psi(x) = \Psi'(x) = \frac{1}{\sqrt{x(K-x)}}, \tag{12}$$

we have

$$|h(x)| = \frac{\psi(S(x)) |S'(x)|}{\psi(x)} = \frac{\sqrt{x(K-x)} |S'(x)|}{\sqrt{S(x)(K-S(x))}} = h_1(x) h_2(x). \tag{13}$$

Note that condition (iii) holds when $\inf |h| = \inf h_1 h_2 > 1$. We check that h_1 and h_2 are C^1 functions. Indeed, the function h_1 can only have singularities at the points $x = 0$ and $x = K$, and since $S(0) = S(K) = 0$ and $S'(0) \neq 0$, $S'(K) \neq 0$, the function S can be represented as the product $S(x) = x(K-x)s_1(x)$, where s_1 is a strictly positive C^1 function, so $h_1(x) = 1/\sqrt{s_1(x)}$ is a C^1 function. The square of the function h_2 is the quotient of the functions S'^2 and $K - S$ and can only have a singularity at the point $\tilde{x}$. Since $S'(\tilde{x}) = 0$, we can represent the function S' as $S'(x) = S''(\tilde{x})(x - \tilde{x})s_2(x)$, where s_2 is a C^1 function. Since $\tilde{x}$ is the zero of the function $K - S$ and its first derivative, the function $K - S$ can be written in the form $K - S(x) = \frac{1}{2}S''(\tilde{x})(x - \tilde{x})^2 s_3(x)$, where again s_3 is a C^1 function different from zero. We can see that

$$h_2^2(x) = \frac{[S'(x)]^2}{K - S(x)} = 2S''(\tilde{x})\frac{s_2^2(x)}{s_3(x)}$$

and $s_2(x) > 0$ for $x \in [0, K]$, so h_2 is a C^1 function. Since $1/\psi(x) = \sqrt{x(K - x)}$ is a continuous function, the condition (iv') holds.

Since S is a unimodal function, conditions (i) and (ii) follow immediately from (9). If f^* is the invariant density of the mapping $\tilde{S}$, then

$$g^*(x) = f^*(\Psi(x))|\Psi'(x)| = \frac{1}{\sqrt{x(K - x)}} f^*(\arccos(1 - 2x/K))$$

is the invariant density of S. Thus g^* is a continuous function in the interval $(0, K)$ and satisfies (11). □

Example 5 Let us consider the transformation $S(x) = cx(1 - x^2)$ defined on the interval $[0, 1]$. Since $S'(x) = c(1 - 3x^2)$, we find that $\tilde{x} = \frac{\sqrt{3}}{3}$ and from the equation $S(\tilde{x}) = 1$ that $c = \frac{3\sqrt{3}}{2}$. The polynomial $1 - S(x) = 1 - cx(1 - x^2)$ has a double zero at $\tilde{x}$, so its decomposition into factors is of the form:

$$1 - S(x) = c\left(x - \frac{\sqrt{3}}{3}\right)^2\left(x + \frac{2\sqrt{3}}{3}\right).$$

Thus, we easily determine h and check the inequality $\inf |h| > 1$:

$$|h(x)| = \frac{\sqrt{x(1 - x)}\,|S'(x)|}{\sqrt{S(x)(1 - S(x))}} = \frac{c|1 - \sqrt{3}x|(1 + \sqrt{3}x)}{\sqrt{c^2(1 + x)\left(x - \frac{\sqrt{3}}{3}\right)^2\left(x + \frac{2\sqrt{3}}{3}\right)}}$$

$$= \frac{3\left(x + \frac{\sqrt{3}}{3}\right)}{\sqrt{(x + 1)\left(x + \frac{2\sqrt{3}}{3}\right)}} \geq \frac{3\frac{\sqrt{3}}{3}}{\sqrt{\frac{2\sqrt{3}}{3}}} = \frac{3}{\sqrt{2\sqrt{3}}} > 1.$$

Example 6 Consider the *Beverton–Holt transformation*

$$S(x) = -ax + \frac{bx}{1 + x}.$$

Observe that $S(K) = 0$ if $b = (K + 1)a$. Since $S'(x) = -a + b/(1 + x)^2$, the function S has the maximum at $\tilde{x} = \sqrt{b/a} - 1 = \sqrt{K + 1} - 1$. Hence

$$S(\tilde{x}) = -a\tilde{x} + \frac{a(K + 1)\tilde{x}}{1 + \tilde{x}} = -a\tilde{x} + a\tilde{x}\sqrt{K + 1} = a(\sqrt{K + 1} - 1)^2,$$

and since $S(\tilde{x}) = K$, we have

$$a = \frac{K}{(\sqrt{K + 1} - 1)^2}, \quad b = \frac{K(K + 1)}{(\sqrt{K + 1} - 1)^2}.$$

Then

$$h_1(x) = \sqrt{\frac{x(K - x)}{S(x)}} = \sqrt{\frac{K - x}{-a + \frac{b}{1+x}}} = \sqrt{\frac{(K - x)(1 + x)}{-a(1 + x) + a(K + 1)}} = \sqrt{\frac{1 + x}{a}},$$

$$h_2(x) = \frac{|S'(x)|}{\sqrt{K - S(x)}} = \frac{\left|-a + \frac{b}{(1+x)^2}\right|}{\sqrt{K + ax - \frac{bx}{1+x}}} = \frac{a\left|(x + 1)^2 - \frac{b}{a}\right|(x + 1)^{-3/2}}{\sqrt{(ax + K)(x + 1) - bx}}$$

$$= \frac{a\left|(x - \tilde{x})(x + 1 + \sqrt{K + 1})\right|(x + 1)^{-3/2}}{\sqrt{a(x - \tilde{x})^2}}$$

$$= \sqrt{a}\left(x + 1 + \sqrt{K + 1}\right)(x + 1)^{-3/2},$$

hence

$$h_1(x)h_2(x) = \frac{x + 1 + \sqrt{K + 1}}{x + 1} \geq \frac{K + 1 + \sqrt{K + 1}}{K + 1} > 1.$$

In some examples it is difficult to find the greatest lower bound of the function h_2. If $\inf |S''| > 0$, then we can estimate h_2 from below by applying Cauchy's mean value theorem: if functions f and g are continuous in the interval $[a, b]$ and differentiable in the interval (a, b), then there is a point $c \in (a, b)$ such that

$$(f(b) - f(a))g'(c) = (g(b) - g(a))f'(c). \tag{14}$$

Let $f(x) = S'^2(x)$, $g(x) = K - S(x)$ and we choose $a = \tilde{x}$, $b = x$, if $x > \tilde{x}$, and $b = \tilde{x}$, $a = x$, if $x < \tilde{x}$. In both cases

$$\frac{S'^2(x)}{K - S(x)} = \frac{2S''(c)S'(c)}{-S'(c)} = -2S''(c)$$

for some $c \in [0, K]$. Thus $h_2 \geq \min \sqrt{2|S''|}$.

Example 7 Consider the *Ricker transformation*

$$S(x) = -ax + axe^{\lambda(K-x)}.$$

It satisfies the condition $S(0) = S(K) = 0$. We change coordinates linearly to get the maximum of S at $\tilde{x} = 1$, and then we find the relations between a, K, and λ. Since $S(1) = K$ and $S'(1) = 0$, we obtain

$$K = 1 - \lambda^{-1}\ln(1 - \lambda), \quad a = \frac{(1-\lambda)K}{\lambda}.$$

Let us observe that

$$h_1(x) = \sqrt{\frac{x(K-x)}{S(x)}} = \sqrt{\frac{K-x}{ae^{\lambda(K-x)} - a}} \geq \sqrt{\frac{K}{ae^{\lambda K} - a}},$$

$$S''(x) = \lambda a(\lambda x - 2)e^{\lambda(K-x)}.$$

Assume that $\lambda K < 2$. Then S'' is a negative and increasing function. Hence

$$|S''(x)| \geq \lambda a |\lambda K - 2|,$$

$$h_2(x) \geq \sqrt{2\lambda a(2 - \lambda K)}.$$

From these inequalities we obtain

$$|h(x)| \geq \sqrt{\frac{2\lambda K(2 - \lambda K)}{e^{\lambda K} - 1}}.$$

Let $c_0 > 0$ be a constant such that $2c_0(2 - c_0) = e^{c_0} - 1$. Then $c_0 \approx 1.0928$ and $\min |h| > 1$ for $\lambda K < c_0$. Since $\lambda K = \lambda - \ln(1 - \lambda)$, the inequality

$$\lambda - \ln(1 - \lambda) < c_0$$

implies that $\min |h| > 1$. We can now choose $\lambda_0 \approx 0.4658$ such that $\min h > 1$ for $\lambda < \lambda_0$.

4   Invariant Measures and Chaos of Structured Populations

Structured populations are usually described by partial differential equations. The dynamical systems generated by such equations are defined on infinite dimensional spaces. As an example we consider the following equation

$$\frac{\partial u}{\partial t} + x\frac{\partial u}{\partial x} = \lambda u, \quad x \in [0, 1], \quad t \geq 0, \quad \lambda > 0 \tag{15}$$

with the initial condition $u(0, x) = v(x)$. This equation generates a semiflow $\{S^t\}_{t \geq 0}$ on the space $X = \{v \in C[0, 1]: v(0) = 0\}$ given by the formula $S^t v(x) = e^{\lambda t} v(e^{-t} x)$. One method of study of ergodic properties of this semiflow developed by Lasota [10] is based on the Krylov–Bogolyubov theorem on the existence of invariant measures for dynamical systems on compact spaces. Unfortunately an invariant measure constructed by this method does not have sufficiently strong ergodic and analytic properties.

Another method developed in the papers [11–13] is based on the following observation. Let $Q: X \to C[0, \infty)$ be the map given by $(Qv)(t) = S^t v(1) = e^{\lambda t} v(e^{-t})$ and let $\{T^t\}_{t \geq 0}$ be the left-side shift on the space $Y = Q(X)$ defined by $(T^t \varphi)(s) = \varphi(s + t)$. Then $T^t \circ Q = Q \circ S^t$ for $t \geq 0$. Let $\xi_t, t \geq 0$, be a stationary stochastic process defined on a probability space (Ω, Σ, P) whose sample paths are in Y. Then the measure $m(A) = P(\omega: \xi_{\cdot}(\omega) \in A)$ defined on the Borel σ-algebra $\mathcal{B}(Y)$ is invariant with respect to $\{S^t\}_{t \geq 0}$. Since dynamical systems $\{S^t\}_{t \geq 0}$ and $\{T^t\}_{t \geq 0}$ are isomorphic the measure $\mu(A) = m(Q(A))$ is invariant with respect to $\{S^t\}_{t \geq 0}$.

Let $w_t, t \geq 0$, be a Wiener process starting from 0 with continuous sample paths and let $\xi_t = e^{\lambda t} w_{e^{-2\lambda t}}$ for $t \geq 0$. Then $\xi_t, t \geq 0$, is a Gaussian stationary process with continuous paths. If m is the measure induced by the process (ξ_t), then the measure μ is induced by the process $\zeta_x = w_{x^{2\lambda}}, x \in [0, 1]$. Since the Wiener measure is positive on nonempty open subsets of X the same property has the measure μ.

Now we check that the dynamical system $(X, \mathcal{B}(X), \mu, S^t)$ is exact. Denote by $\mathcal{F}_T$ the σ-algebra of events generated by the process w_t for $t \in T$. From the definition of the dynamical system $\{S^t\}_{t \geq 0}$ it follows that $S^{-t}(\mathcal{B}(X)) \subseteq \mathcal{F}_{[0, e^{-2\lambda t}]}$ for $t \geq 0$. According to Blumenthal's zero-one law the σ-algebra $\mathcal{F}_{0+} = \bigcap_{s > 0} \mathcal{F}_{[0, s]}$ contains only sets of measure zero, the same property has the σ-algebra $\bigcap_{t \geq 0} S^{-t}(\mathcal{B}(X))$, which proves that the dynamical system $(X, \mathcal{B}(X), \mu, S^t)$ is exact.

In summary, the dynamical system $\{S^t\}_{t \geq 0}$ has an invariant measure, which is positive on nonempty open sets (property (P)) and this system is exact. These properties imply strong chaotic behaviour of this dynamical system. From (P) and ergodicity it follows that μ-almost all trajectories are dense. From (P) and from the mixing property it follows a strong version of instability called *sensitive dependence on initial conditions*: that there exists a constant $\eta > 0$ such that for each point $v \in X$ and for each $\varepsilon > 0$ there exist a point $\bar{v}$ and $t > 0$, such that $\|v - \bar{v}\| < \varepsilon$ and $\|S^t v - S^t \bar{v}\| > \eta$.

Our system has also turbulent trajectories in the sense of Bass. We recall that a trajectory $O(v) = \{S^t v: t \geq 0\}$ of $v \in X$ is *turbulent in the sense of Bass* [14] if there exists an $v_0 \in X$ and a function $\gamma \in C([0, \infty), X)$ such that

(I) $\lim_{T \to \infty} \frac{1}{T} \int_0^T S^t v \, dt = v_0,$

(II) $\lim_{T \to \infty} \frac{1}{T} \int_0^T (S^t v - v_0)(S^{t+\tau} v - v_0) \, dt = \gamma(\tau),$

(III) $\gamma(0) \neq 0$ and $\lim_{\tau \to \infty} \gamma(\tau) = 0$.

The number $\gamma(\tau)$ describes the correlation between the trajectory $O(v)$ and its τ-time shift $O(S^\tau v)$. Condition $\lim_{\tau \to \infty} \gamma(\tau) = 0$ can be interpreted as the "lack of memory" because the trajectory and its τ-time shift are "almost independent" for large τ. The proof of this property is more advanced [13]. We need to check that the second moment of μ is finite, i.e. $\int_X \|v\|^2 \mu(dv) < \infty$ and then apply an ergodic theorem for Banach space valued random variables to prove (I) and (II), and mixing property to prove (III).

We can consider a dynamical system $\{S^t\}_{t \geq 0}$ restricted to the set $X_+ = \{v \in X : v \geq 0\}$. Then the measure μ is induced by the process $\zeta_x = |w_{x^{2\lambda}}|$, $x \in [0, 1]$, is invariant with respect to $\{S^t\}_{t \geq 0}$, positive on nonempty open sets and the system $(X_+, \mathcal{B}(X_+), \mu, S^t)$ is exact. It means that this system has all above mentioned chaotic properties.

Now we apply this observation to study two models of hematopoietic system. The proper functioning of the hematopoietic system largely depends on the process of multiplication and differentiation (maturation) of erythrocytes. The process of erythrocyte production itself is quite complicated, as the formation of a mature erythrocyte occurs as a result of multiple division and differentiation of erythroid stem cells. This process is regulated by the concentration of erythropoietin. It turns out that when the external regulatory mechanisms of the hematopoietic system do not work, the system can act chaotically.

Although our models are rather simple in comparison with other models for erythroid production (e.g. [15, 16]), they are based on the same continuous maturation-proliferation scheme. In both models we assume that the level of morphological development, briefly referred as the cell *maturity*, is expressed by a parameter x in the interval $[0, 1]$. We assume that maturity is a real number $x \in [0, 1]$. The function $u(t, x)$ describes the distribution of cells with respect to their maturity. Assume that maturity grows according to the equation $x' = g(x) > 0$, $g(0) = 0$. We assume that at the time of division, the daughter cells have the same maturity as the parent cell. When the cell reaches maturity $x = 1$ it leaves the bone marrow. In the first model, we assume that each cell splits with rate c or dies with rate d. Let $c = b - d$. Then the distribution $u(t, x)$ of the population at time t relative to maturity x satisfies the equation

$$\frac{\partial u}{\partial t} + \frac{\partial}{\partial x}(g(x)u) = cu. \tag{16}$$

This model can be found in the work [15]. We will consider here a simplified version with $g(x) = x$. Then Eq. (16) can be written in the form of Eq. (15) with $\lambda = c - 1$. Thus if $c > 1$ this model is chaotic.

Now we consider the second model [17]. We assume that when one cell reaches the maturity $x = 1$ it leaves the bone marrow, and then one of cells from the bone marrow splits. This cell is chosen randomly according to the distribution given by the density $p(t, x)$ of all stem cells. Let D_0 be the subset of densities p such that $\int_0^\varepsilon p(x)\,dx > 0$ for each $\varepsilon > 0$. We need to assume that the initial density

$p_0(x) = p(0, x)$ belongs to D_0. Otherwise, the stem cell population will die out in finite time. Then for each $t > 0$ also the density $p(t, x)$ belongs to D_0.

If the density p_0 is a differentiable function, then p satisfies a nonlinear partial differential equation

$$\frac{\partial p}{\partial t} + \frac{\partial}{\partial x}(xp) = p(t, 1)p(t, x). \tag{17}$$

The difference between the Eqs. (16) and (17) is that in the first model, the growth rate is c, while in the second model this rate is equal to $u(t, 1)$. This form of the rate comes from the fact that $u(t, 1)$ describes the number of cells leaving the bone marrow per unit time and this is also how many new cells appears. Is not difficult to check that

$$p(t, x) = \frac{p_0(e^{-t}x)}{\int_0^1 p_0(e^{-t}x)\,dx} \tag{18}$$

and the formula (18) defines a dynamical system $\{P^t\}_{t \geq 0}$ on D_0 given by $P^t p_0(x) = p(t, x)$.

In [17] it is proved that there exists a probability measure ν on $\mathcal{B}(D_0)$ such that ν is positive on nonempty open subsets of D_0 and the dynamical system $(D_0, \mathcal{B}(D_0), \nu, P^t)$ is exact. The proof of this results runs as follows. First we extend the dynamical system $\{S^t\}_{t \geq 0}$ from X_+ onto the space $L^1_+ = \{v \in L^1[0, 1] : v \geq 0\}$ and define the measure $\tilde{\mu}$ by $\tilde{\mu}(A) = \mu(A \cap X_+)$ for $A \in \mathcal{B}(L^1_+)$, where μ is the previously defined invariant measure with respect to $\{S^t\}_{t \geq 0}$. Let Y_0 be the set which consists of all functions $f \in L^1_+$ such that $\int_0^\varepsilon f(x)\,dx > 0$ for each $\varepsilon > 0$. Then we prove that $\tilde{\mu}(Y_0) = 1$, the measure $\tilde{\mu}$ is positive on nonempty open subsets of D_0 and the system $(Y_0, \mathcal{B}(Y_0), \tilde{\nu}, S^t)$ is exact. Next we check that the function $H : X_0 \to D_0$ defined by $Hv = \frac{v}{\|v\|}$ is continuous and satisfies $P^t \circ H = H \circ S^t$ for $t \geq 0$. Let $\nu(A) = \tilde{\mu}(H^{-1}(A))$ for $A \in \mathcal{B}(D_0)$. Then the dynamical system $(D_0, \mathcal{B}(D_0), \nu, P^t)$ has all the required properties.

The method for studying chaos presented above can be extended to a very broad class of partial differential equations of first order with one dimensional space variable x and for some hyperbolic equations. In [18] we study the dynamical system obtained by solving the equation

$$\frac{\partial u}{\partial t} + \frac{\partial}{\partial x}(gxu) = -(m + d)u(t, x) + 4du(t, 2x) \tag{19}$$

which models the size structured cellular population. We show that the considered dynamical system with an appropriately chosen invariant measure m on the space of initial functions X is mixing and the measure m is positive on nonempty open subsets of X.

A major challenge is to study the ergodic properties of dynamical systems generated by partial differential equations with a multidimensional spatial coordinate x. We now consider a model of this type given by the following equation

$$\frac{\partial u}{\partial t}(t, x) + \text{div}(a(x)u(t, x)) = \lambda(1 - u/K(x))u. \tag{20}$$

This equation describes the logistic growth of a structured population. Any individual is characterized by a vector $x \in \mathbb{R}^d$, for example x can be a location in the space or other parameters as age, maturity, size, etc. These parameters change according to the equation $x' = a(x)$. The function $g(x, u) = \lambda(1 - u/K(x))u$ describes the population growth per unit time and $u(t, x)$ is the population density, that is, $\int_A u(t, x)\, dx$ is the number of individuals with parameters in the set A at time t. The positive function K is the local carrying capacity and $\lambda > 0$ is the maximum growth rate.

If $f(x, u) = g(x, u) - u\,\text{div}\,a(x)$, then Eq. (20) can be written in the form

$$\frac{\partial u}{\partial t}(t, x) + a_1(x)\frac{\partial u}{\partial x_1}(t, x) + \cdots + a_d(x)\frac{\partial u}{\partial x_d}(t, x) = f(x, u(t, x)), \tag{21}$$

$t \geq 0$, $x \in D$, where D is a bounded set in $\mathbb{R}^d$ diffeomorphic with the ball $B = \{x \in \mathbb{R}^d : |x| \leq 1\}$. Ergodic properties of the dynamical system generated by this equation are studied in [4].

We assume $a : D \to \mathbb{R}^d$ be a C^1 function, $a(\mathbf{0}) = \mathbf{0}$. Consider the initial problem

$$x'(t) = -a(x(t)), \quad x(0) = x_0 \in D. \tag{22}$$

Denote by $\pi_t x_0$ the solution of (22) and assume that $\lim_{t \to \infty} \pi_t x_0 = \mathbf{0}$ for all $x_0 \in D$. We also assume that $f : D \times \mathbb{R} \to \mathbb{R}$ is a C^1 function. We assume that there exist positive constants A, B such that $f(x, u)u \leq A + Bu^2$ for $(x, u) \in D \times \mathbb{R}$. Then for each C^1 function $v : D \to \mathbb{R}$ there exists a unique solution of (21) satisfying the initial condition $u(0, x) = v(x)$. We define the map $S^t v(x) = u(t, x)$ and extend it to the dynamical system $\{S^t\}_{t \geq 0}$ on the space $C(D)$.

We can restrict the dynamical system $\{S^t\}_{t \geq 0}$ to some subspace of $C(D)$. For our purposes we assume additionally that $f(x, 0) = 0$ for $x \in D$, $\frac{\partial f}{\partial u}(\mathbf{0}, 0) > 0$, and there exists $u_0 > 0$ such that $f(\mathbf{0}, u_0) = 0$, $\frac{\partial f}{\partial u}(\mathbf{0}, u_0) < 0$, $f(\mathbf{0}, u) > 0$ for $u \in (0, u_0)$. Then there exists a unique stationary solution u^+ of (21) such that $u^+(\mathbf{0}) = u_0$. Let

$$X = \{v \in C(D) : v(\mathbf{0}) = 0, \quad v \geq 0, \quad v(x) < u^+(x) \text{ for } x \in D\}.$$

Then the dynamical system $\{S^t\}_{t \geq 0}$ can be restricted to the set X. One of the results of the paper [4] is that there exists a probability measure μ on $\mathcal{B}(X)$ such that μ is positive on nonempty open subsets of X and the dynamical system $(X, \mathcal{B}(X), \mu, S^t)$ is exact. In order to apply this result to Eq. (20) it is sufficient to assume that $\lambda > \operatorname{div} a(\mathbf{0})$.

The main differences in the proofs of these results for one-dimensional and multidimensional x variable are the following. In the construction of the invariant measure μ instead of the Wiener process, we use the random field called the Lévy d-parameter Brownian motion. Instead of the left-side shift on the space $C[0, \infty)$ we use the dynamical system $\{T^t\}_{t \geq 0}$ on the space $C([0, \infty) \times \partial B)$ defined by $T^t w(s, y) = w(s + t, y)$, for $s, t \geq 0$ and $y \in \partial B$, where ∂B is a unit sphere in $\mathbb{R}^d$.

Acknowledgments This research was partially supported by the National Science Centre (Poland) Grant No. 2017/27/B/ST1/00100.

Competing Interests The authors have no conflicts of interest to declare that are relevant to the content of this chapter.

References

1. Mackey, M.C., Glass, L.: Oscillation and chaos in physiological control systems. Science **197**(4300), 287–289 (1977). https://doi.org/10.1126/science.267326
2. Lasota, A.: Stable and chaotic solutions of a first order partial differential equation. Nonlinear Anal. Theory Methods Appl. **5**(11), 1181–1193 (1981). https://doi.org/10.1016/0362-546X(81)90012-2
3. Lasota, A., Mackey, M.C.: Chaos, Fractals and Noise. Stochastic Aspects of Dynamics, vol. 97. Springer, New York (1994). https://doi.org/10.1007/978-1-4612-4286-4
4. Rudnicki, R.: Ergodic properties of a semilinear partial differential equation. J. Differ. Equ. **372**, 235–253 (2023). https://doi.org/10.1016/j.jde.2023.06.046
5. Rudnicki, R.: Stochastic Operators and Semigroups and their Applications in Physics and Biology, Banasiak, J., Mokhtar-Kharroubi, M. (eds.), pp. 255–318. Springer, Cham (2015). https://doi.org/10.1007/978-3-319-11322-7_6
6. Rudnicki, R., Tyran-Kamińska, M.: Piecewise Deterministic Processes in Biological Models. SpringerBriefs in Applied Sciences and Technology, Mathematical Methods. Springer, Cham (2017). https://doi.org/10.1007/978-3-319-61295-9
7. Boyarsky, A., Góra, P.: Laws of Chaos: Invariant Measures and Dynamical Systems in One Dimension. Probability and Its Applications. Birkhäuser, Boston (1997). https://doi.org/10.1007/978-1-4612-2024-4
8. Lasota, A., Yorke, J.A.: Exact dynamical systems and the Frobenius. Perron operator. Trans. Am. Math. Soc. **273**, 375–384 (1982). https://doi.org/10.2307/1999212
9. Lasota, A., Yorke, J.A.: On the existence of invariant measures for piecewise monotonic transformations. Trans. Am. Math. Soc. **186**, 481–488 (1973). https://doi.org/10.2307/1996575
10. Lasota, A.: Invariant measures and a linear model of turbulence. Rendiconti Semin. Mat. della Univ. Padova **61**, 39–48 (1979)
11. Brunovskỳ, P., Komornik, J.: Ergodicity and exactness of the shift on $C[0, \infty)$ and the semiflow of a first-order partial differential equation. J. Math. Anal. Appl. **104**(1), 235–245 (1984). https://doi.org/10.1016/0022-247X(84)90045-3
12. Rudnicki, R.: Invariant measures for the flow of a first order partial differential equation. Ergodic Theory Dyn. Syst. **5**(3), 437–443 (1985). https://doi.org/10.1017/S0143385700003059

13. Rudnicki, R.: Strong ergodic properties of a first-order partial differential equation. J. Math. Anal. Appl. **133**(1), 14–26 (1988). https://doi.org/10.1016/0022-247X(88)90361-7
14. Bass, J.: Stationary functions and their applications to the theory of turbulence: I. Stationary functions. J. Math. Anal. Appl. **47**(2), 354–399 (1974). https://doi.org/10.1016/0022-247X(74)90026-2
15. Lasota, A., Mackey, M.C., Ważewska-Czyżewska, M.: Minimizing therapeutically induced anemia. J. Math. Biol. **13**, 149–158 (1981). https://doi.org/10.1007/BF00275210
16. Mackey, M.C., Dörmer, P.: Continuous maturation of proliferating erythroid precursors. Cell Prolifer. **15**(4), 381–392 (1982). https://doi.org/10.1111/j.1365-2184.1982.tb01055.x
17. Rudnicki, R.: Chaoticity of the blood cell production system. Chaos Interdiscip. J. Nonlinear Sci. **19**(4), 043112 (2009). https://doi.org/10.1063/1.3258364
18. Rudnicki, R.: Chaoticity and invariant measures for a cell population model. J. Math. Anal. Appl. **393**(1), 151–165 (2012). https://doi.org/10.1016/j.jmaa.2012.03.055

Quantitative Insights into Glucose Regulation: A Review of Mathematical Modeling Efforts

Moisés Santillán (iD)

Abstract Mathematical modeling has proven to be an effective method for studying glucose homeostasis. Over the last several decades, quantitative models have provided critical insights into the complex regulatory mechanisms underlying optimal blood glucose control. This review summarizes significant advances in elucidating the dynamic glucose-insulin relationship from the molecular to the organismal levels using mathematical representations. We show how early conceptual models paved the way for increasingly sophisticated multi-scale computational models. We also talk about how the ongoing collaboration between modeling and experiments helps us understand glucose homeostasis and drives therapeutic innovation. Finally, we discuss new frontiers such as integrating glucose regulation into whole-body regulatory networks and unraveling long-term compensatory mechanisms.

1 Introduction

Homeostasis refers to the ability of an organism or system to maintain internal stability and equilibrium despite fluctuations in the external environment. The concept of homeostasis is fundamental to physiology and mathematical modeling of biological systems. The pioneering work of Claude Bernard in the nineteenth century laid the foundations for the modern understanding of homeostasis. Bernard introduced the concept of the milieu interieur or internal environment, and recognized that complex organisms possess an intrinsic ability to regulate and maintain the stability of this internal environment [1]. Walter Cannon expanded on Bernard's ideas in the twentieth century, providing a comprehensive theory of homeostasis focused on the wisdom of the body in maintaining steady states. Cannon coined the term homeostasis and defined it as the coordinated physiological processes which maintain most of the steady states in the organism [2].

M. Santillán (✉)
Center for Research and Acvanced Studies, Apodaca, México
e-mail: msantillan@cinvestav.mx

Y. Mori et al. (eds.), *Dynamics of Physiological Control*, Lecture Notes on Mathematical Modelling in the Life Sciences,
https://doi.org/10.1007/978-3-031-82396-1_7

A key development was the application of mathematical control theory and cybernetics to biological systems by Wiener and Rosenblueth in the 1940s. They proposed that feedback control systems actively maintain the constancy of the internal environment in the face of external disturbance [3]. This provided a quantitative framework for understanding homeostasis through systems dynamics concepts. Today, homeostasis remains a core paradigm in physiology, centered on the idea that organisms regulate internal variables within a narrow range around a target setpoint.

Glucose homeostasis represents a prime example for exploring homeostasis through mathematical modeling. Plasma glucose levels are maintained within a narrow range despite variances in glucose intake and utilization. This is accomplished through an elegant control system integrating the endocrine pancreas, liver, and peripheral tissues. Mathematical models have proven valuable for teasing apart the complexity of this vital homeostatic mechanism. Recent decades have witnessed extensive modeling efforts to simulate glucose-insulin dynamics and gain insights into diabetes pathology [4, 5]. In summary, the journey from Bernard's milieu interieur to today's computational models of glucose homeostasis epitomizes the growth of systems thinking in physiology. This review will chart the key milestones in this intellectual progression and synthesize current modeling approaches for glucose homeostasis.

2 The Glucose-Insulin Regulatory System

Our current understanding of the physiological mechanisms regulating glucose homeostasis stands on the shoulders of seminal discoveries in the nineteenth and early twentieth centuries. In mid nineteenth century, Claude Bernard elucidated the function of the liver in maintaining euglycemia through dynamic storage and release of glucose from glycogen [6]. This work foreshadowed Bernard's paradigm-shifting conception of homeostasis. In 1869, Paul Langerhans discovered the pancreatic islets housing the insulin-producing beta cells that bear his name, the Islets of Langerhans [7].

The critical link between the pancreas and glucose metabolism was established in 1889 when Minkowski and Mering induced diabetes by removing the pancreas in dogs [8]. This paved the way for Banting, Best and Macleod's Nobel-awarded isolation of insulin in 1921 at the University of Toronto and demonstration of its glucose-lowering effects in diabetic dogs [9]. The clinical preparation of insulin enabled insulin replacement therapy, transforming type 1 diabetes from a deadly disease to a manageable condition.

These foundational discoveries instantiated the framework for investigating the physiological machinery governing glucose homeostasis. In subsequent decades, intensive research has elucidated the intricate multisystem regulation of glucose metabolism. Our current understanding reveals a sophisticated neuroendocrine control system integrating the pancreas, liver, fat tissue, gastrointestinal system and brain. Below, this clockwork is briefly reviewed, primarily based on the classic physiology textbook by Guyton [10].

When plasma glucose level is viewed as a flux balance system, it is clear that it is affected by the following processes: glucose influx from carbohydrate ingestion, endogenous glucose production, insulin-stimulated glucose uptake, and insulin-independent glucose uptake.

Endogenous glucose production (EGP) encompasses the processes of glycogenolysis (the breakdown of glycogen into glucose) and gluconeogenesis (the production of glucose from non-carbohydrate carbon substrates) that occur in the liver. One of insulin's primary metabolic effects is the suppression of endogenous glucose production. Evidence suggests that the majority of insulin's suppression of EGP is mediated by an extra-hepatic, indirect mechanism (single gateway hypothesis) [11].

Glucose uptake is carried out by a family of sugar transporter proteins collectively termed as GLUTs [12, 13], through ATP-independent facilitated diffusion [13–16]. There are over 10 different types of glucose membrane transporters. Of them, the most significant for glucose homeostasis are GLUTs 1–4 [14]. Of these, GLUT1 and GLUT3 have a high affinity for glucose (K_M = 1.4 mM). GLUT2 transporters have a low affinity for glucose (K_M = 17 mM). This contrasts with GLUT4 medium range affinity (K_M = 5 mM), which lies just above normoglycemic levels.

The major cellular mechanism that decreases blood glucose after carbohydrates ingestion is insulin-stimulated glucose uptake into various tissues; remarkably, skeletal muscle. GLUT4, which is highly expressed in adipose tissue and skeletal muscle, is mostly located in intracellular pools in the non-stimulated state, and is acutely redistributed to the plasma membrane in response to insulin and other stimuli. More specifically, GLUT4 is subject to a continuous recycling pathway in which exocytosis is tightly regulated [13].

GLUTs 1–3 operate in a constitutive fashion and thus are responsible for insulin-independent glucose uptake. GLUT1 and GLUT3 are responsible for maintaining a basal rate of insulin-independent glucose uptake into erythrocytes, endothelial cells and neurons. GLUT2 are located in the plasma membrane of hepatocytes and pancreatic β cells, thus mediating a small fraction of overall glucose uptake which mostly serves sensing purposes for these specialized cells. and mediates insulin-stimulated glucose uptake.

Insulin dynamics is governed by by insulin secretion and clearance. According to Kalwat and Cobb [17], pancreatic islet β cells secrete insulin in a biphasic manner in response to glucose stimulation. Glucose sensing by β cells increases intracellular ATP levels and causes an influx of calcium (Ca^{2+}) ions. Elevated calcium concentrations near the plasma membrane cause insulin secretion in two phases: an initial high rate within minutes of glucose stimulation and a slow, sustained release lasting longer than 30 minutes. In the initial phase, insulin granules already docked at the cell membrane are released via exocytosis. Calcium also causes a translocation of reserve granules within the cell towards the plasma membrane for release in the second, sustained phase of secretion [18]. This second,

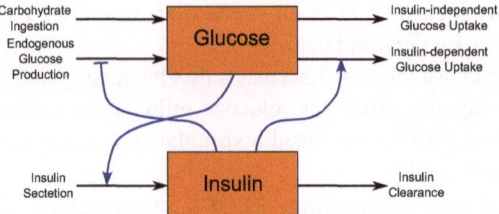

Fig. 1 Schematic representation of the glucose-homeostasis regulatory pathway. Boxes represent plasma levels of glucose and insulin. Black arrows represent input and output fluxes from the corresponding boxes. Blue lines denote regulation of specific fluxes by either glucose or insulin. Arrowheads correspond to upregulation, while hammerheads represent downregulation

slower process involves reorganisation of cortical filamentous actin cytoskeleton [19–21]. Finally, The liver is the primary site of insulin clearance, although kidneys and muscles also participate [22].

The mechanisms described above reveal a couple of negative feedback loops underlying the glucose homeostasis regulatory pathway (see Fig. 1). Specifically, rising plasma glucose levels boost insulin secretion, while rising insulin levels suppress endogenous glucose production and boost insulin-dependent glucose update.

The complex physiological machinery that governs glucose homeostasis provides a fascinating testbed for mathematical modeling approaches. Researchers have used quantitative tools to elucidate glucose-insulin dynamics since Bergman's seminal minimal model in the 1970s. Following Bergman's foundation, extensive modeling efforts have been undertaken to further investigate the complexities of glucose regulation using mathematical representations fitted to experimental data. These efforts have incorporated additional layers of regulation, recognizing that glucose levels are controlled not just by the negative feedback loop of insulin and glucose depicted in Fig. 1, but through additional regulatory loops involving hormones like glucagon, secreted by the pancreas in response to low blood sugar levels, and glucocorticoids, stress hormones that increase glycogenolysis and gluconeogenesis. Capturing these interactions as well as those between glucose control and other physiological systems regulating appetite, food intake, and circadian rhythms would require expanding Fig. 1 to include further regulatory loops. Advancing such multi-loop models necessitates more sophisticated mathematical representations. But first, Bergman's seminal minimal model, which pioneered the use of modeling in diabetes research, will be reviewed.

One of the most influential mathematical models of glucose regulation is the minimal model developed by Richard Bergman and colleagues in the late 1970s [23]. This elegant model emerged from efforts to find a simple yet physiologically-based model that could capture key aspects of glucose homeostasis using data from intravenous glucose tolerance tests (IVGTT).

In an IVGTT, a bolus of glucose is administered intravenously, bypassing gastrointestinal absorption variables. This allows for a direct assessment of the body's insulin response independent of gut-derived factors. Studies using the minimal model have effectively utilized the automated and reproducible glycemic spikes induced by IVGTTs to discern physiological parameters like insulin sensitivity that quantify core aspects of glucose disposal. The absence of gastrointestinal influences also simplifies modeling of the IVGTT profile compared to an oral challenge. Furthermore, IVGTTs probe the beta cell's acute insulin secretory capacity in isolation, separate from its response to post-prandial incretins. This functionality has been critically important for mathematical models aiming to disentangle the relative roles of insulin resistance versus secretion in glucose dysregulation.

Bergman's minimal model consists of just two differential equations—Eqs. (1) and (2)—representing the dynamics of plasma glucose level and insulin in a remote interstitial fluid compartment (IFC). A key insight was that insulin's effects on glucose metabolism were delayed, requiring insulin to enter the remote compartment before acting. Parameter S_G, which Bergman et al. named "glucose effectiveness," acounts for insulin independent uptake and endogenous glucose production. Variable $I(t)$ accounts for plasma insulin level. Parameters p_2 and p_3 are transport coefficients for insulin diffusing from plasma to the IFC and vice versa. From these, Bergman et al. defined an insulin sensitive parameter as $S_I = p_3/p_2$.

$$\frac{dG}{dt} = -[S_G + X(t)]\,G(t), \tag{1}$$

$$\frac{dX}{dt} = p_2 I(t) - p_3 X(t). \tag{2}$$

By fitting the minimal model to IVGTT glucose ($G(t)$) and insulin data ($I(t)$), clinically useful indices like S_I and S_G could be estimated for an individual. S_I was rigorously validated against the glucose clamp technique and found to be an accurate index of insulin sensitivity [24]. The model further enabled assessing the relative contributions of insulin resistance versus β-cell dysfunction in diabetes by computing a Disposition Index as the product of S_I and the amplitude of the acute insulin secretion in response to glucose injection [25].

In the original model, the delay in insulin action was hypothesized to be due to slow transport of insulin from plasma into interstitial fluid bathing insulin-responsive tissues like muscle. This conceptual model was supported by careful studies measuring insulin dynamics across capillaries, confirming the rate-limiting role of insulin transport in its action on glucose metabolism [26].

The minimal model proved useful for comparing insulin sensitivity between individuals and populations. For example, it revealed lower insulin sensitivity in obese subjects and enabled tracking changes with weight loss interventions [27]. However, it was found that the model underestimated insulin sensitivity in very resistant individuals with inadequate insulin secretion. This limitation led to modifications like adding tolbutamide or insulin injection during IVGTTs to boost insulin profiles.

While the minimal model was developed using IVGTT data, subsequent studies extended it to analyze data from oral glucose tolerance tests (OGTT). In an OGTT, glucose is administered orally to induce changes in blood glucose and insulin levels over the subsequent hours. This real-world challenge to the regulatory system elicits information about how the various control mechanisms act in an integrated manner in response to nutrient ingestion. Features such as the glucose curve shape, insulinemic response profile, and timing of peaks provide quantitative readouts of the functional interplay between tissues. The non-steady state nature of the OGTT also allows models to account for time delays and distributed mechanisms in a way more representative of physiological reality compared to intravenous challenges.

The minimal model parameters have also been incorporated into more extensive models of glucose-insulin dynamics and pancreatic insulin secretion. However, the original model remains widely used today due to its simplicity and clinical utility.

The minimal model has shed light on determinants of glucose tolerance beyond just insulin sensitivity. For instance, studies revealed the importance of hepatic insulin clearance, which can vary substantially between individuals and impact insulin levels and action [28]. Glucose effectiveness has also been identified as a modulator of diabetes risk, with Bergman proposing it as a second defense protecting those with insulin deficiency [29].

Other modeling approaches have been developed to complement the minimal model, providing additional perspectives on glucose regulation. For example, the oral minimal model analyzes OGTT data to assess insulin secretion and sensitivity [30]. Nevertheless, Bergman's original model retains unique advantages in clinical applications due to its simplicity and interpretability.

In summary, the minimal model provided a parsimonious yet powerful mathematical representation of glucose homeostasis. Its impact demonstrates the synergistic value of mathematical modeling paired with experimental studies for advancing scientific understanding. The model remains an influential tool in diabetes research and clinical practice.

3   Further Modeling Approaches and Applications

The minimal model developed by Bergman and colleagues paved the way for countless subsequent efforts to model glucose dynamics using mathematical approaches. Since it was first introduced, researchers have built upon its foundation to create more detailed mechanistic models, incorporate additional hormones and processes, and apply modeling to address a diverse range of questions related to glucose homeostasis. This section will synthesize some of the developments in mathematical modeling that have emerged in recent years, highlighting their novel contributions and clinical or biological insights gained.

3.1 Diabetes Diagnosis, Progression and Treatment

Mathematical models of glucose dynamics can provide insights into specific aspects of diabetes diagnosis, progression, and treatment. By quantitatively representing key pathological features, modeling approaches allow simulation of disease timecourses and evaluation of potential intervention effects. However, this introductory subsection takes a broad scope, surveying examples that address diverse clinical questions rather than focusing on any single area. Papers were selected to demonstrate how modeling has been applied to elucidate facets of diabetes pathogenesis, identify diagnostic biomarkers, and design improved therapies. This wide range is addressed through discrete examples rather than an exhaustive review. The following models were chosen to represent this breadth of topics investigated through mathematical modeling in diabetes research.

Dalla Man et al. [31] validate a previously introduced oral minimal model (OMM) [32] against the reference tracer method for estimating the rate of glucose appearance from a meal (R_{meal}) and insulin sensitivity (SI) following an oral glucose load. The reference method is a two-step procedure where a two-tracer protocol and non-steady state glucose kinetics model are first used to reconstruct R_{meal}. The reconstructed R_{meal} is then input into the classic single-compartment minimal model to estimate SI. This reference method provides an accurate determination of R_{meal} and SI but requires radioactive tracers. The OMM couples the standard minimal model ordinary differential equation (ODE) system with a parametric description of R_{meal} as a piecewise linear function. This allows the OMM to simultaneously estimate R_{meal} and SI from plasma glucose and insulin data, avoiding tracer use. The authors compare OMM and reference method estimates of R_{meal} and SI using data from 88 subjects who underwent the reference triple-tracer meal protocol. Good agreement was observed between R_{meal} profiles and average SI estimates. However, limitations were noted for individual predictions due to assumptions required to fix OMM parameters.

Ha et al. [33] developed a mathematical model of the pathogenesis, prevention, and reversal of type 2 diabetes (T2D) using a system of ordinary differential equations (ODEs). The model builds on an existing negative feedback model of the insulin-glucose regulatory system by Topp et al. [34] that incorporates the dynamic regulation of beta-cell mass. The authors extended this framework by making two aspects of beta-cell function dynamic on intermediate timescales: (1) the glucose dose-response curve (DRC), which is shifted leftwards under persistent hyperglycemia, and (2) the maximum insulin secretion rate, which increases to amplify the secretory response. These changes in function act in parallel with the long-term growth of beta-cell mass to compensate for insulin resistance. The model is able to reproduce patterns of dysregulation seen in animal models of T2D, such as hyperinsulinemia preceding hyperglycemia as observed in rats fed high-fat diets. It provides quantitative insight into the relative roles of insulin resistance versus secretion defects in diabetes progression. Key to the model behavior is the existence of a threshold glucose level separating stable normoglycemic and hyperglycemic states. The paper discusses how this threshold framework explains phenomena like the difficulty of curing diabetes compared to prevention, and the superior efficacy of bariatric surgery versus dietary interventions. Overall, the ODE model captures important dynamics of beta-cell compensation for insulin resistance and failure of this process leading to T2D. It provides a useful theoretical framework for understanding disease mechanisms.

Pérez-Rivera et al. [35] developed a computational mathematical model of glucose-insulin dynamics and the progression of diabetes mellitus. They formulated a system of ordinary differential equations to represent the interactions between glucose, insulin, pancreatic beta cells, and fat deposition over time. This constitutes a dynamical systems model approach. The differential equations capture the effects of insulin resistance due to increasing fat deposition, modeled via a logistic function, as well as beta cell death due to prolonged exposure to high glucose levels (glucose toxicity). Through simulations, the model demonstrates how an unhealthy high-fat diet can lead to hyperglycemia and eventually insulin deficiency through increasing resistance and decreasing beta cell mass, reflecting the progression from type 2 to type 1 diabetes. The authors suggest future work could involve better characterizing fat consumption patterns, incorporating exercise effects, and statistically linking the model predictions to clinical data to enable patient-specific prognosis and treatment planning.

Contreras et al. [36] introduced a mathematical model consisting of delay differential equations (DDE) to represent glucose-insulin dynamics during an oral glucose tolerance test (OGTT). DDEs were employed for the first time in an OGTT-related model, introducing delays to account for transit between compartments in the gastrointestinal tract. The model was applied to fit data from a cohort of 407 clinically healthy patients who underwent a standard 5-point OGTT. The DDE model could fit the diverse glucose-insulin profiles observed in the cohort, including hypoglycemic individuals, single/double peak curves, etc. This suggests the DDE model can capture different physiological mechanisms underlying glycemic control. The authors introduced a new strategy to solve the parameter fitting problem, using information from clinical records and the OGTT itself to constrain and reshape the feasible parameter space, allowing robust estimation of physiologically meaningful parameters for each patient. This was confirmed by means of a parameter sensitivity analysis. From their results, the authors argue that their approach could allow personalized diagnosis and treatment.

Lombarte et al. [37] developed linear ordinary differential equation (ODE) models to study glucose homeostasis in rats with type 1 and type 2 diabetes mellitus (DMT1 and DMT2). The models consist of a system of 3–5 linear ODEs representing the changes in plasma glucose, plasma insulin, glucose in the digestive system, and glucose in urine over time. The models have a small number of physiological parameters that can be estimated for each individual rat using measurements of plasma glucose and insulin levels in response to an oral glucose tolerance test or subcutaneous insulin injection. This allows quantifying the different processes involved in glucose regulation, such as insulin secretion, insulin clearance, glucose uptake by tissues, liver glucose metabolism, and renal glucose excretion, for each rat. The authors validated the models by showing good agreement between model simulations and experimentally measured glucose and insulin dynamics. The DMT1 model was then applied to study the effects of fluoride exposure on parameters related to insulin secretion, insulin sensitivity, and liver function. The results from the model agreed with previous experimental findings. A parameter representing the insulin secretion rate was also evaluated as a potential diagnostic test for low insulin secretion in DMT1 rats, showing high sensitivity and specificity compared to the standard HOMA-IR test. HOMA-IR stands for "Homeostatic Model Assessment of Insulin Resistance". It is a method used to estimate insulin resistance and beta-cell function. Specifically: HOMA-IR is computed using the formula:

$$HOMA\text{-}IR = (Fasting\ glucose \times Fasting\ insulin)/22.5.$$

A higher HOMA-IR value indicates higher insulin resistance.

López-Palau et al. [38] developed what they call a physiological-based pharmacokinetic-pharmacodynamic (PB-PKPD) mathematical model to characterize blood glucose dynamics in patients with type 2 diabetes mellitus (T2DM). The model consisted of 28 nonlinear ordinary differential equations and was based on a previous PB-PKPD model for healthy humans by Alvehag and Martin [39]. However, López-Palau et al. modified certain metabolic functions to capture the pathophysiology of T2DM, including impaired insulin-mediated glucose uptake, excessive hepatic glucose production, and pancreatic beta-cell dysfunction. Parameters governing these "sensitive" metabolic rates were calibrated by fitting the model to clinical data from T2DM patients using both static and dynamical approaches. The PB-PKPD model could successfully simulate blood glucose responses to different stimuli, such as intravenous glucose infusion and oral glucose tolerance tests. Simulation results aligned reasonably well with clinical measurements from T2DM patients. The authors proposed that their PB-PKPD model could prove useful for developing and testing controllers to regulate blood glucose in T2DM. Additionally, linking the model with pharmacokinetic-pharmacodynamic characterizations of anti-diabetic drugs may provide insights into personalized diabetes treatment.

Ng et al. [40] developed a minimal mathematical model of blood glucose homeostasis consisting of a differential equation for glucose levels and a closed proportional-integral control loop to model the aggregate effects of insulin, glucagon, etc. It has the advantage of having only 3 tunable parameters: proportional control, integral control, and time scale of feedback delays. They showed that this parsimonious model has a unique, stable equilibrium solution within normal glucose range for healthy parameters. The authors further found that the model fits well Continuous Glucose Monitor (CGM) data from 3 different studies on healthy individuals. The fitted parameter values were consistent across studies, falling into normal distributions, suggesting that the model assesses glucose regulation in a robust, standardized way. From this, the authors propose that the model could potentially be used with CGM data for quick, non-invasive diagnosis of prediabetes by detecting changes in the model parameters from their normal distributions.

Brenner et al. [41] estimated the parameters of a quasi-linear mathematical model (consisting of 3 coupled ODEs) of glucose-insulin homeostasis introduced by Lombarte et al. [42] by fitting the results of intraperitoneal glucose tolerance tests performed in lean (fed a normal diet) and obese mice (fed a high-fat diet for 3 months). They found that 5 out of the 9 model parameters were significantly different between the groups: basal insulin levels were higher in obese mice, indicating hyperinsulinemia. The rate constant for insulin-dependent glucose uptake was lower in obese mice, suggesting insulin resistance. The rate constant for insulin-independent uptake was higher in obese mice, likely due to increased body mass. The rate constant for liver glucose transfer was lower in obese mice, indicating insulin resistance in the liver. The insulin level where the liver switches from glucose release to uptake was higher in obese mice, also suggesting insulin resistance in the liver. From this, the authors conclude that the model can be used as a tool to study alterations in glucose-insulin homeostasis parameters with obesity progression.

Bizzoto et al. [43] collected data from 123 subjects across 5 different tests: clamp studies (euglycemic clamp maintains blood glucose at basal levels through a variable glucose infusion, hyperglycemic clamp maintains elevated blood glucose levels), oral glucose tolerance tests (OGTT) which measure blood glucose and insulin responses to a prescribed glucose drink, and meal tests which measure responses to a mixed meal. Tracer and insulin data allowed modeling glucose uptake kinetics using a nonlinear system of ordinary differential equations (ODEs). The authors developed a mathematical model of glucose kinetics based on a circulatory model with two compartments (heart-lung and peripheral tissues) interconnected. Glucose uptake in the peripheral tissues was represented in the model as a saturable Michaelis-Menten function of the glucose concentration at the site of action, with the maximum uptake rate being an insulin-dependent Hill function of the insulin concentration at the site of action. Both glucose and insulin concentrations at the site of action in the model were described using distributed-delay compartment models to account for transport delays. The single mathematical model, composed of these nonlinear ODEs, could then accurately describe the glucose kinetics in all tests. The model predicted glucose clearance suppression at higher glucose levels. Furthermore, model simulations showed that glucose uptake saturation significantly increased glucose levels during an OGTT and decreased levels during insulin infusion, compared to a linear model. From this, the authors argue that glucose uptake saturation is an important physiological phenomenon that impacts glucose tolerance and should be considered when modeling glucose homeostasis.

Dietrich et al. [44] developed a time-discrete nonlinear feedback model based on the MiMe-NoCoDI modeling platform for endocrine systems [45]. The model employed physiological concepts such as saturation kinetics and non-competitive inhibition to describe the insulin-glucose feedback loop. Two novel parameters were derived from the model: SPINA-GBeta, estimating the secretory capacity of beta cells, and SPINA-GR, estimating insulin receptor gain. These parameters were validated by re-analyzing data from two clinical cohorts. SPINA-GBeta and SPINA-GR were found to correlate with measures of glucose metabolism and body composition in both cohorts. SPINA-GR was lower in subjects with diabetes or prediabetes. A hyperbolic relationship was observed between SPINA-GBeta and SPINA-GR in healthy subjects, suggesting beta cell compensation for declining insulin sensitivity, which is impaired in prediabetes/diabetes. According to the authors, the model provides a theoretical basis for a simple diagnostic test using just fasting glucose and insulin concentrations. If validated further, it could potentially be useful for screening purposes and clinical research.

Murillo et al. [46] introduced a mathematical model of delay differential equations to study the dynamics of glucose, insulin, and free fatty acids (FFA) and how FFA may impact insulin resistance progression. The model incorporates an explicit time delay for insulin secretion in response to glucose levels. This allows the model to better capture the physiological delays in insulin secretion compared to prior "minimal models" formulated as systems of ordinary differential equations. The delay differential equation model describes the interactions between glucose, insulin stimulated by both glucose and FFA levels with a time delay, and FFA production regulated by insulin. The authors fitted the model to clinical data from patients with varying degrees of metabolic health, from normal to prediabetic to diabetic, who underwent bariatric surgery. The results showed improvements in insulin sensitivity, glucose effectiveness, and other metabolic parameters after bariatric surgery compared to before surgery across patient groups, matching previous clinical findings that bariatric surgery can improve insulin regulation of glucose and FFA through weight/fat loss. The authors concluded that their delay differential equation model provides a quantitative framework to study the impact of FFA dysregulation on insulin resistance progression and the benefits of bariatric surgery.

Salentine et al. [47] developed an ordinary differential equation (ODE) model to study the competition between cancer cells and healthy glucose-dependent cells for a shared glucose resource. The model considers the supply and consumption of glucose by healthy tissues and cancer cells, as well as the suppression of healthy tissue glucose uptake by cancer cells. According to their results from simulation studies, cancer was found to erupt earlier in diabetic patients compared to non-diabetic patients due to their reduced glucose intake. Additionally, cancer aggressiveness was shown to increase with higher cancer glucose consumption rate and efficiency of glucose use by cancer cells in the model. Simulating anti-diabetic drugs like metformin in the model delayed cancer eruption by increasing healthy tissue glucose uptake. However, reducing glucose intake alone only worked if cancer uptake also declined in the model. The model predictions regarding tumor growth rates, body weight loss, and blood glucose levels were found to reasonably match clinical observations in cancer patients. The authors propose that their ODE model could be further adapted to specific cancer types by parameterizing it using PET scan data on glucose consumption rates. They argue that the model provides a framework to test interventions aimed at controlling systemic glucose levels in cancer patients.

Strilka et al. [48] employed a mathematical model to simulate the effects of subcutaneous regular insulin and lispro insulin (a fast-acting insulin analog) injections on stress hyperglycemia in critically ill patients receiving continuous enteral nutrition. The model used was a physiologically-based mathematical model, consisting of a set of delay differential equations, that described the glucose-insulin system. It included seven compartments representing glucose distribution in the body, insulin distribution and action, and enteral nutrition absorption. The model simulated and compared the effects of the two types of insulin on mean glucose levels and glucose variability. The results suggested that regular insulin tended to decrease mean glucose and glucose variability more linearly compared to lispro. This makes regular insulin better suited for sliding scale protocols. Lispro tended to increase glucose variability even if it decreased mean glucose, sometimes in a nonlinear manner, suggesting that lispro may not be optimal for continually postprandial patients. Higher lispro doses produced hypoglycemia when nutrition was discontinued, whereas regular insulin did not. Severe insulin resistance seemed to predict when the glucose-insulin system was more sensitive to changes in insulin levels. This could result in "rebound hyperglycemia" and increased variability with lispro. The study highlights the importance of considering nutrition status when designing insulin protocols.

3.2   Detailed Modeling

While simple models like Bergman's minimum model have been extremely helpful, there has been a growing interest in recent years for developing more intricate, mechanistic models of glucose homeostasis. By utilizing the abundance of experimental data now accessible on intracellular signaling pathways, tissue-specific kinetics, and organ interaction, these models seek to integrate increased biological realism. Despite inherent difficulties in testing and interpreting complicated models, their ability to recreate various aspects of glucose management enables important discoveries that simple models are unable to capture. Here, we discuss a specific instance of the research value of thorough quantitative representations of glucose homeostasis.

Herrgårdh et al. [49] introduced an updated multi-level mathematical model of glucose homeostasis in humans. This is a physiologically-based mathematical model that integrates data across multiple scales. The model employs systems of ordinary differential equations (ODEs) to describe the dynamic interactions between components. Key improvements include more accurate organ glucose uptake proportions, tissue-specific timing, and blood flow effects. The model integrates data across multiple scales, from intracellular signaling pathways described by systems of ODEs, to whole-body glucose fluxes between organs also modeled using ODEs. The modular structure, with separable subsystems for each organ, allows integration of future intracellular data described by additional ODEs into the whole-body model. The authors incorporated timing differences in glucose uptake between muscle and adipose tissue, by modeling intracellular glucose phosphorylation in adipocytes using a system of ODEs. This results in an earlier peak uptake in adipose tissue compared to muscle. The model also accounts for the impact of blood flow on insulin-stimulated glucose uptake in adipose tissue: increased blood flow amplifies the effect of insulin on glucose uptake, modeled through additional ODEs describing the relationship. The model was able to fit over 300 data points across more than 40 time series and dose-response curves, describing intracellular processes, organ fluxes, and whole-body meal responses under different perturbations. The authors argue that this multi-level mathematical model provides a useful platform for integrating future experimental data and gaining systems-level insights into glucose regulation.

3.3   Interaction with Other Hormones and Metabolites

Glucose homeostasis is intricately intertwined with numerous other regulatory processes in the body. Mathematical modeling provides a means to elucidate these complex interactions by representing glucose-insulin dynamics along with other influential hormones, nutrients, and signaling pathways within an integrated computational framework. Recent modeling efforts have explored the interplay of glucose regulation with diverse factors including amino acids, fatty acids, leptin, and glucagon. This subsection summarizes illustrative examples of models that encapsulate the interconnected nature of glucose control with other molecular and physiological determinants.

Kadota et al. [50] developed a mathematical model to investigate the potential of leptin as a treatment for type 1 diabetes. Leptin is a hormone released by adipose (fat) tissue that plays a key role in regulating appetite and metabolism. The model integrates a brain-centered glucoregulatory system (BCGS) where leptin plays a key role into conventional models of insulin-glucose dynamics. The model employs a system of ordinary differential equations to represent the different components involved. It includes equations to model: (1) plasma and tissue glucose concentrations and their interaction with insulin and other factors like cortisol, (2) insulin secretion and dynamics in the blood and liver, (3) leptin synthesis in adipose tissue and its effects in the brain and peripheral tissues, (4) glucose uptake and production rates in the liver and tissues, (5) glucose transport across the blood-brain barrier, and (6) administration and absorption of exogenous insulin. The model represents both the insulin-dependent and independent effects of leptin on glucose regulation. Leptin is shown to lower blood glucose through inhibiting cortisol production and increasing tissue insulin sensitivity. The model was validated by comparing simulations to experimental data on leptin's glucose-lowering effects in mice and rats. In silico experiments using the model demonstrate that adding leptin treatment improves glucose control and reduces insulin requirements compared to insulin monotherapy in type 1 diabetes patients. This suggests that leptin may help overcome limitations of current diabetes treatments and has potential benefits when added to treatment regimens. The model provides a means to study leptin's mechanisms and efficacy without human risks.

Morettini et al. [51] introduced a mathematical compartmental model to assess kinetics of glucagon and its inhibition by insulin in individuals using OGTT data. The model consists of a system of linear ordinary differential equations (ODEs) with two compartments: a well-mixed plasma glucagon compartment and a remote C-peptide compartment. Glucagon is secreted by pancreatic alpha cells and plays an important role in glucose homeostasis by stimulating hepatic glucose production and opposing the effects of insulin. The model aims to assess the sensitivity of glucagon inhibition to glucose-induced insulin secretion, referred to as "alpha-cell insulin sensitivity" and denoted as SGLUCA(t). SGLUCA(t) is modeled as a time-varying parameter governing glucagon secretion, while glucagon and C-peptide clearance rates are constant parameters. The model was validated using mean experimental data from 17 subjects undergoing a 5-hour OGTT, providing good fits. Further validation on 100 virtual subjects showed the model could reliably reproduce different glucagon curves. SGLUCA(t) showed significant negative correlation with the ratio of area under the curve (AUC) of remote C-peptide to AUC of glucagon. Estimation of SGLUCA(t) was consistent between 5-hour and 3-hour OGTT data, but not 2-hour. In summary, this simple compartmental ODE model allowed individual-level estimation of glucagon kinetics parameters and alpha-cell insulin sensitivity from OGTT data.

Subramanian et al. [52] developed a delay differential equation based model to analyze data from an isoglycemic intravenous glucose infusion (IIGI) experiment. In an IIGI experiment, glucose is infused intravenously while carefully adjusting the rate to match plasma glucose excursions seen with an oral glucose tolerance test, without stimulating gut hormone release. The developed model extends previous glucose-insulin models to include glucagon dynamics using a system of three coupled delay differential equations for glucose, insulin, and glucagon. The model was fit simultaneously to IIGI experimental data from individuals both with and without type 2 diabetes. This allowed estimation of parameters related to insulin sensitivity, glucagon action, and glucose-dependent insulin and glucagon secretion dynamics free of gut-related effects. The paper finds impairments in these parameters in T2D and observes correlations with diabetes markers. This phenomenological model provides insights into the roles of alpha- and beta-cell dysfunction in T2D pathophysiology using a minimal model approach.

Morettini et al. [53] developed a mathematical model to describe the effects of amino acids (AAs) on insulin secretion and kinetics during a mixed meal tolerance test (MMTT). They proposed and compared five different ordinary differential equation (ODE) models using previously published experimental data on insulin, glucose and branched-chain amino acid (BCAA) levels in healthy and type 2 diabetes subjects during a MMTT. The selected best model was a simple non-compartmental ODE model with a single first-order differential equation relating changes in insulin concentration to plasma glucose concentration (k_{GL}), plasma BCAA concentration (k_{AA}), basal insulin secretion rate (BRI) and constant insulin clearance rate (k_I). In healthy subjects, the model parameter k_{AA} representing the sensitivity of insulin secretion to BCAAs was estimated to be positive, indicating a stimulatory effect of BCAAs on insulin secretion. In type 2 diabetes patients,

k_{AA} was estimated not significantly different from zero, suggesting no significant effect of BCAAs on insulin secretion. The model was further validated on simulated data for 100 healthy and 100 T2D virtual subjects, confirming the findings. The authors suggest their simple ODE model could help describe individual amino acid effects on insulin secretion under different metabolic conditions, aiding dietary optimization for diabetes management.

Hampton et al. [54] developed a differential equations-based mathematical model to describe the interactions between glycerol and insulin dynamics during an OGTT in adolescent girls with obesity. Glycerol is a marker of lipolysis, the breakdown of fats in adipose tissue to release free fatty acids and glycerol. For glycerol dynamics, they developed an explicit ordinary differential equation model representing the suppressive effects of insulin action on lipolysis, modeled as a decreasing Hill function of insulin action. They also used an existing oral minimal model, a system of ordinary differential equations describing how insulin action enhances glucose uptake and reduces endogenous glucose production. They compared the dynamics of insulin action on glycerol (representing adipose tissue metabolism and lipolysis) versus insulin action on glucose (representing muscle and liver metabolism) using the glycerol model and an oral minimal model. Their results showed that insulin action on glycerol/adipose tissue, suppressing lipolysis and glycerol release, peaked earlier and more closely followed insulin concentrations compared to insulin action on glucose/muscle and liver, enhancing glucose uptake. This suggests tissue-specific differences in insulin action dynamics, with a more rapid insulin effect on adipose tissue compared to muscle and liver. The differences in insulin action dynamics may reflect the extreme insulin resistance and compensatory hyperinsulinemia present in this adolescent population with obesity. The findings highlight the importance of using mathematical modeling approaches to gain insights into tissue-specific metabolic dysregulation in high-risk youth populations.

3.4   Oscillatory Dynamics

A fascinating dynamical phenomenon that has been observed experimentally in glucose homeostasis is the presence of oscillatory patterns in insulin, glucose, and other metabolites. Ultradian oscillations with periods on the order of about 50–200 minutes have been measured in humans and animal models. In addition, circadian rhythms modulating glucose regulation over the 24-hour sleep-wake cycle have been identified. Mathematical modeling has provided a useful approach to investigate this complex dynamic behavior. This subsection examines modeling studies that have harnessed the power of quantitative tools to elucidate the genesis of oscillations in glucose homeostasis across multiple timescales.

Ultradian oscillations with periods of 50–200 minutes have been observed experimentally in insulin secretion and blood glucose levels in humans and animals. Sturis et al. [55] developed a system of nonlinear ordinary differential equations (ODEs) model of causal loop diagrams and differential equations to investigate whether the oscillations could arise from the feedback loops between glucose and insulin. The model includes two major negative feedback loops: (1) glucose stimulates insulin secretion, and insulin inhibits glucose production (2) glucose stimulates insulin secretion, and insulin increases glucose utilization. Simulations of the ODE model reproduced key features of the ultradian oscillations observed experimentally, including self-sustained oscillations during constant glucose infusion and damped oscillations after glucose ingestion. The oscillations were found to depend critically on the time delay in insulin's effect on glucose production and the two-compartment distribution of insulin in the ODE model. The results suggest the ultradian oscillations can arise entirely from the feedback loops between glucose and insulin in the ODE model, without requiring an intrinsic pancreatic pacemaker.

Tolic et al. [56] developed a mathematical model of the insulin-glucose feedback system in humans based on nonlinear differential equations. The model consists of a system of delay differential equations representing the dynamics of insulin secretion from the pancreas, transport between plasma and intercellular space, and effects of insulin on hepatic glucose production and peripheral glucose utilization in tissues. A key feature of the model is that it exhibits self-sustained ultradian oscillations in insulin and glucose levels similar to oscillations observed experimentally, suggesting the oscillations may arise from an inherent oscillatory mechanism or instability in the nonlinear insulin-glucose feedback loop. The authors used the model to investigate why oscillatory insulin delivery has a greater hypoglycemic effect compared to constant delivery at the same mean rate, as observed in experiments. Analysis of the nonlinear differential equation model and simulations suggest the enhanced effect of oscillatory insulin is primarily due to nonlinear effects on suppressing hepatic glucose production. Specifically, when insulin levels are below the inflection point of the dose-response curve for hepatic glucose production, oscillations extend the time spent at higher insulin levels, enhancing the overall suppression of glucose production. The model was later improved by Tolic et al. by incorporating additional nonlinear terms representing receptor dynamics, effects of hyperglycemia, and other refinements. However, the central result from simulations, that oscillatory insulin reduces mean glucose levels predominantly by acting on hepatic glucose production, remained qualitatively unchanged in the improved nonlinear delay differential equation model.

Woller and Gonze [57] studied the effect of restricting food intake to the normal resting phase in mice using a mathematical modeling approach. They developed an ordinary differential equation model to describe the interactions between components of the circadian clock and glucose metabolism in mouse pancreatic beta cells. The model incorporated light/SCN cues, nutrient cues, positive and negative feedback loops in the circadian clock, and the regulation of insulin secretion. The SCN, or suprachiasmatic nucleus, is a region of the brain that acts as the central circadian pacemaker, entraining peripheral clocks via neural and hormonal signals. The model consisted of 13 ordinary differential equations the accounted for two embedded networks: a metabolic module describing glucose-insulin dynamics, and a circadian oscillator module representing the core circadian clock feedback loops in beta cells. When simulating restricted daytime feeding by shifting the food intake signal while keeping SCN cues unchanged, the model showed differential phase shifts in clock gene expression. Some genes shifted by about 12 hours to follow the inverted food cues, while others only partially shifted due to the conflicting unchanged SCN cues. This differential phase shift in the model caused a misalignment between nutrient cues and clock-controlled cues that regulate insulin exocytosis. The model predicted that this subsequently leads to hypoinsulinemia, hyperglycemia, and loss of food anticipatory behavior experimentally observed in mice under these feeding conditions. The study proposed that conflicting inputs to local circadian clocks from light cycles versus food intake disrupts the normal phase relation between clock components in the model. This then cascades to disrupt downstream metabolic processes regulated by the clock, such as insulin secretion.

Hara and Satake [58] developed an ordinary differential equation (ODE) model to examine how circadian regulation of glucose metabolism (specifically glycogen and triglyceride production/breakdown) can minimize risks of energy depletion and hyperglycemia. The model consisted of a system of ODEs describing the dynamics of glucose, glycogen, and triglyceride levels in the liver under different feeding schedules. By running simulations across different peak times for glycogen and triglyceride synthesis, the model found the optimal peak times that minimized each risk. In both cases, glycogen synthesis should peak at the end of the active period. However, triglyceride synthesis should peak at the end of the resting period to minimize energy depletion risk, and peak in the middle of the active period to minimize hyperglycemia risk. The model showed that increased fat accumulation from eating during the resting period (as seen in shift workers) emerges from circadian regulation optimized to prevent energy depletion, but not hyperglycemia. This suggested that circadian coordination of metabolism in mammals may be adapted to minimize energy depletion risk, and increased fat accumulation in irregular schedules is a byproduct of this optimization. The study proposes that the mismatch between this optimization and modern lifestyles/eating habits may contribute to increased obesity and metabolic disease risk.

3.5 *Bistable Behavior*

Bistability is an intriguing dynamical phenomenon whereby a system can switch between two distinct stable states yet persist in either state over time. This on/off toggle property has been proposed to play an important role in glucose homeostasis by conferring switch-like control of insulin signaling and glucose uptake. Some modeling studies have suggested that representing insulin-responsive cells or tissues as bistable systems could help explain experimental findings related to all-or-none insulin activation, reactive hypoglycemia, and nonlinear whole-body dose responses. This subsection surveys modeling efforts that have specifically investigated bistable responses as a key mode of regulation in glucose homeostasis.

Wang [59] used a mathematical modeling approach, consisting of a system of ordinary differential equations, to propose an "adjustable threshold hypothesis" to explain insulin resistance and glucose-insulin homeostasis. The main ideas are: individual cells respond to insulin in an all-or-none manner, switching between no glucose uptake and maximal uptake at threshold insulin concentrations Ion and Ioff; this creates hysteresis: a delayed switch-on to spare glucose for the brain, and a delayed switch-off to avoid hyperglycemia; cells have heterogeneous thresholds, so their collective response produces a graded dose-response curve for the whole body; insulin resistance arises from elevated thresholds in many cells, requiring higher insulin to trigger glucose uptake. This is adaptive in pregnancy to spare glucose for the fetal brain; hysteresis causes reactive hypoglycemia: when insulin levels fall, cells remain active longer due to delayed switch-off, leading to an undershoot in blood glucose; insulin acts primarily to restrict peripheral glucose utilization, not promote it. It is a glucose stamp to spare glucose for the brain. According to the authors, their hypothesis provides an intuitive dynamical explanation for insulin resistance and glucose homeostasis using a system of ordinary differential equations. It suggests the adjustable thresholds may allow early detection of prediabetes.

Tomar et al. [60] developed an ODE-based mathematical model to study the significance of bistability in metabolic control and the insights provided into disease mechanisms and potential therapeutics. In summary, the model represents the insulin and glucagon signaling pathways that regulate metabolic homeostasis through synchronized activation of anabolic and catabolic processes respectively, using a system of ordinary differential equations (ODEs). Nonlinear positive feedback loops in these pathways can elicit a bistable response in the ODE model, where the network activity toggles between low and high levels. This allows optimal metabolic regulation across different conditions like fasting, feeding, or exercise. Bistability confers benefits like signaling memory, adaptivity, and robust homeostasis. Dysregulation of bistability in the ODE model is linked to metabolic diseases like obesity, diabetes, and non-alcoholic fatty liver disease (NAFLD), which is characterized by excessive fat accumulation in the liver not caused by alcohol. By analysing their ODE-based model for the integrated insulin-glucagon network, the authors showed how bistable response shifts in insulin resistant and glucagon resistant conditions simulated in the model, indicating impaired metabolic regulation. Insulin resistance reduces anabolic efficiency, while glucagon resistance impairs catabolic mobilization of glucose and amino acids in the model. This provides insights on reversing these resistances through lifestyle interventions in the model.

Akhtar et al. [61] present experimental evidence that mouse skeletal muscle cells (C2C12 myotubes) exhibit a hysteretic response to insulin stimulation consistent with bistability. That is, there are two distinct thresholds: an upper threshold where insulin switches the cell on "(activates glucose uptake) and a lower threshold where insulin switches the cell off". Using FRET imaging of myotubes expressing an Akt biosensor, the switch-on threshold (Ion) was typically around 300 pM insulin and the switch-off threshold (Ioff) was around 100 pM. The FRET experiments were done by incrementally increasing insulin doses, reaching saturation, then slowly decreasing insulin back to zero. Akt remained active during the decremental dosing until insulin dropped below Ioff, confirming bistability. Western blotting of myotube populations also provided evidence for bistability, with Akt phosphorylation increasing significantly only above 300 pM insulin. Mathematical modeling showed that a bistable insulin response curve is essential to fit an ODE model describing oral glucose tolerance test data in humans. This suggests the bistable response observed in vitro is physiologically relevant. The bistable model provided insights into metabolic regulation. For example, explaining reactive hypoglycemia and the sigmoid shape of whole-body insulin response curves described by an ODE model. The upper threshold Ion correlates with body fat composition, suggesting it could be a useful biomarker of metabolic health. In summary, the study demonstrates that skeletal muscle cells respond to insulin in a bistable manner and this bistability helps reconcile the competing needs for both rapid and safe glucose regulation described by the ODE model of the body.

3.6   Open Problems and Future Directions

While mathematical modeling has already provided numerous insights into glucose regulation, many open questions remain to be addressed through quantitative approaches. Two key areas are integrating glucose homeostasis with other interconnected systems and elucidating long-term compensatory mechanisms.

An important challenge is representing interactions between glucose-insulin dynamics and other regulatory systems such as the hypothalamic-pituitary-adrenal (HPA) axis and mTOR pathway. The HPA axis regulates glucocorticoid release and stress responses, which impact insulin sensitivity and glucose metabolism [62]. The mTOR pathway acts as a nutrient sensor that controls cell growth and metabolism, responding to insulin, amino acids, and other signals relevant to glucose homeostasis [63]. Mathematical models that integrate glucose-insulin dynamics with HPA axis and mTOR models could provide insights into how these systems work in concert to maintain metabolic regulation.

Another key direction is using modeling to uncover long-term compensatory mechanisms underlying glucose homeostasis. While short-term dynamics have been well-characterized, the longer-term adaptive processes are less understood. For instance, how does β-cell mass expand to compensate for increasing insulin demands? [64] What causes β-cell failure in late-stage diabetes? [65]. Mathematical representations of cellular metabolic sensing pathways, mitochondrial energetics, proliferation signaling, and tissue remodeling could shed light on these compensatory responses. Dynamic modeling can complement experimental work by simulating chronic perturbations such as insulin resistance or hyperglycemia over months-years, accelerating the study of slow adaptive phenomena.

In summary, quantitative modeling provides a powerful approach to tackle complex questions, like the ones referred above, at the whole-body scale, guiding and complementing reductionist experiments. Collaborative efforts between modelers, physiologists and clinicians will be key to realizing the full potential of computational models to elucidate both acute and chronic facets of glucose regulation.

4 Conclusions

Over a century of research has progressively unraveled the intricate physiological mechanisms governing glucose homeostasis. Pivotal discoveries by scientists like Claude Bernard, Paul Langerhans, and Banting and Best laid the foundations for our modern understanding of glucose regulation. The pioneering application of mathematical modeling and cybernetics to biology by Wiener, Rosenblueth, and others enabled a quantitative, systems-level perspective on homeostasis.

Bergman's minimal model catalyzed efforts to quantify glucose-insulin dynamics through mathematical representations fitted to clinical data. This allowed estimation of physiologically meaningful parameters related to insulin sensitivity and secretion, facilitating diabetes research and diagnosis. The model exemplified the synergistic power of combining mathematical and experimental approaches.

In the decades since, modeling glucose homeostasis has continued to rapidly progress on multiple fronts. More detailed mechanistic models have incorporated intracellular pathways, organ glucose fluxes, and tissue-specific kinetics. Novel clinical applications have been suggested, using modeling to guide individualized diabetes prediction and treatment. Interactions between glucose regulation and other systems like amino acid metabolism, circadian rhythms, and hypothalamic-pituitary-adrenal axis have been clarified through quantitative representations. Fundamental dynamical phenomena like bistability and oscillations have been elucidated and connected to whole-body physiology.

The manuscript highlights the diverse array of mathematical techniques that have been applied to model glucose homeostasis. These range from simple models using systems of ordinary differential equations, as in Bergman's seminal minimal model, to more complex formulations involving delay differential equations, nonlinear dynamics, and quasi-linear representations.

Each approach offers advantages for particular research questions. Ordinary differential equation models remain useful for gaining clinical insights due to their relatable parameters, while delay differential equations can capture physiological transport lags more realistically. The choice of modeling formalism depends on the problem scope and resolution required. Simpler models support broad hypothesis testing, whereas intricate models can generate novel mechanistic insights. No single approach suffices. Their synergistic application has propelled understanding, from elucidating minimal regulatory principles to unraveling the interaction with other physiological systems.

Moving forward, integrating glucose homeostasis models with broader regulatory networks and expanding to long-term adaptive behaviors represent exciting frontiers. Collaborations between modelers and experimentalists will remain key to overcoming the inherent complexity of glucose regulation across scales. Mathematical modeling is poised to continue generating fresh insights and guiding therapeutic innovations to improve metabolic health worldwide. The long arc of discovery from Bernard's milieu interieur to today's computational models epitomizes the enduring value of quantitative systems biology approaches in unraveling physiological complexity.

While machine learning techniques have produced impressive results, the role of mechanistic modeling should not be diminished. Data-driven methods can generate highly accurate predictions but have limitations in providing insights into underlying biological mechanisms. Glucose regulation involves complex, nested feedback interactions between multiple organs that emerge from processes operating across various temporal and spatial scales. Mechanistic models formalize current structural and pathway knowledge related to homeostasis based on control theory principles. They allow proposed mechanisms to be interrogated through in silico experimentation, complementing traditional hypothesis-driven research approaches. Even as data volumes continue growing exponentially, developing deeper under-standing will still rely on integrating observations into frameworks of interacting components across different timescales. Mechanistic models offer explanatory power that enhances discovery efforts compared to prediction alone. Used in an integrated manner with machine learning techniques applied to comprehensive datasets, mechanistic modeling continues to hold promise for advancing research beyond prediction to gain new revelations about this vital physiological regulatory system.

Acknowledgments To my mentor and friend, Professor Michael C. Mackey. It is with the utmost gratitude that I dedicate this chapter to you, on the occasion of your 80th birthday. Your guidance, support and wisdom over the years have been truly inspirational. You sparked my interest in mathematical modeling and provided invaluable feedback that helped shape me into the researcher I am today. I am deeply honored to call you my mentor. Thank you for your insights and guidance, but most of all for your kindness and friendship. I wish you all the best. I'm grateful to Laura Sánchez-Gómez for critically reading the manuscript.

Competing Interests The author has no conflicts of interest to declare that are relevant to the content of this chapter.

References

1. Bernard, C.: Lectures on the Phenomena of Life Common to Animals and Plants. American Lecture Series, vol. 1. Thomas, Springfield (1974)
2. Cannon, W.B.: Physiol. Rev. **9**(3), 399 (1929). https://doi.org/10.1152/physrev.1929.9.3.399
3. Wiener, N., Rosenblueth, A.: Arch. Inst. Cardiol. Mexico **16**(3), 205–265 (1946)
4. Bergman, R.N., Ider, Y.Z., Bowden, C.R., Cobelli, C.: Am. J. Physiol. Endocrinol. Metabol. **236**(6), E667 (1979). https://doi.org/10.1152/ajpendo.1979.236.6.e667
5. Cobelli, C., Pacini, G.: Diabetes **37**(2), 223 (1988). https://doi.org/10.2337/diab.37.2.223
6. Young, F.G.: BMJ **1**(5033), 1431 (1957). https://doi.org/10.1136/bmj.1.5033.1431
7. Sakula, A.: J. R. Soc. Med. **81**(7), 414 (1988). https://doi.org/10.1177/014107688808100718

8. Mering, J., Minkowski, O.: Archiv Exper. Pathol. Pharmakol. **26**(5–6), 371 (1890). https://doi.org/10.1007/bf01831214

9. Banting, F.G., Best, C.H., Collip, J.B., Campbell, W.R., Fletcher, A.A.: Diabetes **5**(1), 69 (1956). https://doi.org/10.2337/diab.5.1.69

10. Guyton, A.C., Hall, J.E.: Textbook of Medical Physiology. Guyton Physiology, 11th edn. W B Saunders, London (2005)

11. Anderwald, C., Gastaldelli, A., Tura, A., Krebs, M., Promintzer-Schifferl, M., Kautzky-Willer, A., Stadler, M., DeFronzo, R.A., Pacini, G., Bischof, M.G.: J. Clin. Endocrinol. Metabol. **96**(2), 515 (2011). https://doi.org/10.1210/jc.2010-1398

12. Bogan, J.S.: Annu. Rev. Biochem. **81**(1), 507 (2012). https://doi.org/10.1146/annurev-biochem-060109-094246

13. Huang, S., Czech, M.P.: Cell Metab. **5**(4), 237 (2007). https://doi.org/10.1016/j.cmet.2007.03.006

14. Navale, A.M., Paranjape, A.N.: Biophys. Rev. **8**(1), 5 (2016). https://doi.org/10.1007/s12551-015-0186-2

15. Carruthers, A.: Phys. Rev. **70**(4), 1135 (1990)

16. Chen, L.Y.: Biochim. Biophys. Acta Biom. **1864**(9), 183975 (2022). https://doi.org/10.1016/j.bbamem.2022.183975

17. Kalwat, M.A., Cobb, M.H.: Pharmacol. Therapeut. **179**, 17 (2017). https://doi.org/10.1016/j.pharmthera.2017.05.003

18. Ohara-Imaizumi, M., Fujiwara, T., Nakamichi, Y., Okamura, T., Akimoto, Y., Kawai, J., Matsushima, S., Kawakami, H., Watanabe, T., Akagawa, K., Nagamatsu, S.: J. Cell Biol. **177**(4), 695 (2007). https://doi.org/10.1083/jcb.200608132

19. Kalwat, M.A., Thurmond, D.C.: Exp. Mol. Med. **45**(8), e37 (2013). https://doi.org/10.1038/emm.2013.73

20. Mourad, N.I., Nenquin, M., Henquin, J.C.: Am. J. Physiol. Cell Physiol. **299**(2), C389 (2010). https://doi.org/10.1152/ajpcell.00138.2010

21. Wang, Z., Thurmond, D.C.: J. Cell Sci. **122**(7), 893 (2009). https://doi.org/10.1242/jcs.034355

22. Duckworth, W.C., Bennett, R.G., Hamel, F.G.: Endocrine Rev. **19**(5), 608 (1998). https://doi.org/10.1210/edrv.19.5.0349

23. Bergman, R.N., Ider, Y.Z., Bowden, C.R., Cobelli, C.: Am. J. Physiol. Endocrinol. Metabol. **236**(6), E667 (1979). https://doi.org/10.1152/ajpendo.1979.236.6.e667

24. Bergman, R.N., Prager, R., Volund, A., Olefsky, J.M.: J. Clin. Invest. **79**(3), 790 (1987). https://doi.org/10.1172/jci112886

25. Bergman, R.N., Phillips, L.S., Cobelli, C.: J. Clin. Invest. **68**(6), 1456 (1981). https://doi.org/10.1172/jci110398

26. Yang, Y.J., Hope, I.D., Ader, M., Bergman, R.N.: J. Clin. Invest. **84**(5), 1620 (1989). https://doi.org/10.1172/jci114339

27. Welch, S., Gebhart, S.S.P., Bergman, R.N., Phillips, L.S.: J. Clin. Endocrinol. Metabol. **71**(6), 1508 (1990). https://doi.org/10.1210/jcem-71-6-1508

28. Kim, S.P., Ellmerer, M., Kirkman, E.L., Bergman, R.N.: Am. J. Physiol. Endocrinol. Metabol. **6**, E1581 (2007). https://doi.org/10.1152/ajpendo.00351.2006

29. Bergman, R.N.: Mt. Sinai J. Med. **69**(5), 280 (2002)

30. Breda, E., Cavaghan, M.K., Toffolo, G., Polonsky, K.S., Cobelli, C.: Diabetes **50**(1), 150 (2001). https://doi.org/10.2337/diabetes.50.1.150

31. Man, C.D., Caumo, A., Basu, R., Rizza, R., Toffolo, G., Cobelli, C.: Am. J. Physiol. Endocrinol. Metabol. **287**(4), E637 (2004). https://doi.org/10.1152/ajpendo.00319.2003

32. Man, C.D., Caumo, A., Cobelli, C.: IEEE Trans. Biomed. Eng. **49**(5), 419 (2002). https://doi.org/10.1109/10.995680

33. Ha, J., Satin, L.S., Sherman, A.S.: Endocrinology **157**(2), 624 (2015). https://doi.org/10.1210/en.2015-1564

34. Topp, G., Promislow, K., Devries, G., Miura, R.M., Finegood, D.T.: J. Theor. Biol. **206**(4), 605 (2000). https://doi.org/10.1006/jtbi.2000.2150

35. Pérez-Rivera, D.T., Torres-Torres, V.L., Torres-Colón, A.E., Cruz-Aponte, M.: Math. Biosci. Eng. **13**(5), 1043 (2016). https://doi.org/10.3934/mbe.2016029

36. Contreras, S., Medina-Ortiz, D., Conca, C., Olivera-Nappa, A.: Front. Bioeng. Biotechnol. **8**, 195 (2020). https://doi.org/10.3389/fbioe.2020.00195

37. Lombarte, M., Lupo, M., Brenda, L.F., Campetelli, G., Marilia, A.B., Basualdo, M., Rigalli, A.: J. Theor. Biol. **439**, 205 (2018). https://doi.org/10.1016/j.jtbi.2017.12.001

38. López-Palau, N.E., Olais-Govea, J.M.: Sci. Rep. **10**(1), 12697 (2020). https://doi.org/10.1038/s41598-020-69629-0

39. Alvehag, K., Martin, C.: Proceedings of the 45th IEEE Conference on Decision and Control. IEEE, Piscataway (2006). https://doi.org/10.1109/cdc.2006.377192

40. Ng, E., Kaufman, J.M., van Veen, L., Fossat, Y.: PLOS Digit. Health **1**(7), e0000072 (2022). https://doi.org/10.1371/journal.pdig.0000072

41. Brenner, M., Mohsen Abadi, S.E., Balouchzadeh, R., Lee, H.F., Ko, H.S., Johns, M., Malik, N., Lee, J.J., Kwon, G.: Heliyon **3**(7), e00310 (2017). https://doi.org/10.1016/j.heliyon.2017.e00310

42. Lombarte, M., Lupo, M., Campetelli, G., Basualdo, M., Rigalli, A.: Math. Biosci. **245**(2), 269 (2013). https://doi.org/10.1016/j.mbs.2013.07.017

43. Bizzotto, R., Natali, A., Gastaldelli, A., Muscelli, E., Krssak, M., Brehm, A., Roden, M., Ferrannini, E., Mari, A.: Am. J. Physiol. Endocrinol. Metabol. **311**(2), E346 (2016). https://doi.org/10.1152/ajpendo.00045.2016

44. Dietrich, J.W., Dasgupta, R., Anoop, S., Jebasingh, F., Kurian, M.E., Inbakumari, M., Boehm, B.O., Thomas, N.: Sci. Rep. **12**(1), 17659 (2022). https://doi.org/10.1038/s41598-022-22531-3

45. Dietrich, J.W., Böhm, B.: GMDS 2015; 60. Jahrestagung der Deutschen Gesellschaft für Medizinische Informatik p. Biometrie und Epidemiologie e.V. (GMDS) (2015). https://doi.org/10.3205/15GMDS058

46. Murillo, A.L., Li, J., Castillo-Chavez, C.: Math. Biosci. Eng. **16**(5), 5765 (2019). https://doi.org/10.3934/mbe.2019288

47. Salentine, N., Doria, J., Nguyen, C., Pinter, G., Wang, S.E., Hinow, P.: Bullet. Math. Biol. **85**(58), 1 (2023). https://doi.org/10.1007/s11538-023-01146-3

48. Strilka, R.J., Stull, M.C., Clemens, M.S., McCaver, S.C., Armen, S.B.: Theor. Biol. Med. Model. **13**(3), 3 (2016). https://doi.org/10.1186/s12976-016-0029-2

49. Herrgårdh, T., Li, H., Nyman, E., Cedersund, G.: Front. Physiol. **12**, 619254 (2021). https://doi.org/10.3389/fphys.2021.619254

50. Kadota, R., Sugita, K., Uchida, K., Yamada, H., Yamashita, M., Kimura, H.: J. Theor. Biol. **456**, 213 (2018). https://doi.org/10.1016/j.jtbi.2018.08.008

51. Morettini, M., Burattini, L., G"obl, C., Pacini, G., Ahre'n, B., Tura, A.: Front. Endocrinol. **12**, 611147 (2021). https://doi.org/10.3389/fendo.2021.611147

52. Subramanian, V., Bagger, J.I., Holst, J.J., Knop, F.K., Vilsbøll, T.: Front. Physiol. **13**, 911616 (2022). https://doi.org/10.3389/fphys.2022.911616

53. Morettini, M., Palumbo, M.C., Göbl, C., Burattini, L., Karusheva, Y., Roden, M., Pacini, G., Tura, A.: Front. Endocrinol. **13**, 966305 (2022). https://doi.org/10.3389/fendo.2022.966305

54. Hampton, G.S., Bartlette, K., Nadeau, K.J., Cree-Green, M., Diniz Behn, C.: Front. Physiol. **13**, 895118 (2022). https://doi.org/10.3389/fphys.2022.895118

55. Sturis, J., Polonsky, K.S., Mosekilde, E., Van Cauter, E.: Am. J. Physiol. Endocrinol. Metabol. **260**(5), E801 (1991). https://doi.org/10.1152/ajpendo.1991.260.5.E801

56. Tolić-Nørrelykke, I.M., Mosekilde, E., Sturis, J.: J. Theor. Biol. **207**(3), 361 (2000). https://doi.org/10.1006/jtbi.2000.2180

57. Woller, A., Gonze, D.: Sci. Rep. **8**(1), 13641 (2018). https://doi.org/10.1038/s41598-018-31804-9

58. Hara, A., Satake, A.: J. Math. Biol. **83**(26), 1 (2021). https://doi.org/10.1007/s00285-021-01645-8

59. Wang, G.: J. R. Soc. Interface **11**(101), 20140892 (2014). https://doi.org/10.1098/rsif.2014.0892

60. Tomar, M., Somvanshi, P.R., Kareenhalli, V.: Mol. Biol. Rep. **49**, 5017 (2022). https://doi.org/
    10.1007/s11033-022-07175-w
61. Akhtar, J., Han, Y., Han, S., Lin, W., Cao, C., Ge, R., Babarinde, I.A., Jia, Q., Yuan, Y., Chen,
    G., Zhao, Y., Ye, R., Liu, G., Chen, L., Wang, G.: iScience **25**(12), 105561 (2022). https://doi.
    org/10.1016/j.isci.2022.105561
62. Stephens, M.A.C., Mahon, P.B., McCaul, M.E., Wand, G.S.: Psychoneuroendocrinology **66**,
    47 (2016). https://doi.org/10.1016/j.psyneuen.2015.12.021
63. Kennedy, B.K., Lamming, D.W.: Cell Metabol. **23**(6), 990 (2016). https://doi.org/10.1016/j.
    cmet.2016.05.009
64. Khadra, A., Schnell, S.: Mol. Aspects Med. **42**, 78 (2015). https://doi.org/10.1016/j.mam.2015.
    01.005
65. Cinti, F., Bouchi, R., Kim-Muller, J.Y., Ohmura, Y., Sandoval, P.R., Masini, M., Marselli, L.,
    Suleiman, M., Ratner, L.E., Marchetti, P., Accili, D.: J. Clin. Endocrinol. Metabol. **101**(3),
    1044 (2016). https://doi.org/10.1210/jc.2015-2860

Why Does a Stick Balanced on the Fingertip Fall?

John Milton and Tamás Insperger

Abstract An inverted pendulum can be fully stabilized using time-delayed feedback. In contrast, for human stick balancers the stick always falls. It is suggested that stick falls are a consequence transient micro-chaotic dynamics which naturally arises because of the interplay between time-delayed feedback, a sensory dead zone and the frequency-dependent encoding of force.

1 Introduction

How does a human learn to balance a stick on the fingertip?

This question has fascinated engineers [15, 18–20, 48, 57, 65, 66], mathematicians [9, 28, 52, 53, 55], neuro-scientists [8, 14, 17, 29, 34, 38, 46, 50], and physicists [3–7] for over 80 years. The availability of modern motion capture systems has made it possible to record the fluctuations of the vertical displacement angle of the "balanced" stick in minute detail. Thus it is possible to compare mathematical predictions directly to experimental observations. From the mathematical point of view, the upright position of an inverted pendulum can be fully stabilized by time–delayed feedback [20, 57]. In contrast, even for the most skilled human stick balancer, the stick falls. Why?

Human stick balancing involves a voluntary, goal-directed sequence of movements. Thus stick balancing is a very different motor task from those, such as swallowing, which are largely controlled by hard-wired reflexes and central pattern generators. In other words, stick balancing is a motor skill, not a reflex. This

J. Milton (✉)
Department of Neurology, Dell Medical School, University of Texas at Austin, Austin, TX, USA
e-mail: john.milton@austin.utexas.edu

T. Insperger
Department of Applied Mechanics, Faculty of Mechanical Engineering, Budapest University of Technology and Economics, Budapest, Hungary
e-mail: insperger@mm.bme.hu

© The Author(s), under exclusive license to Springer Nature Switzerland AG 2025
Y. Mori et al. (eds.), *Dynamics of Physiological Control*, Lecture Notes
on Mathematical Modelling in the Life Sciences,
https://doi.org/10.1007/978-3-031-82396-1_8

observation has a number of consequences: (1) practice is required to both attain and maintain the skill level; (2) stick balancing skill is not transferable to other balancing skills, i.e. it exhibits specificity; (3) skill levels differ considerably between individuals; and (4) even for the most skilled individual who has attained an average balance time of minutes, the next balance trial may last only a few seconds! Thus stick balancers encounter the same issues faced by individuals who learn other skilled movements, perhaps even those who try to learn to play golf [41]! A consequence is that it is important to distinguish between those who are at the early stages of "learning a task" (novices) from those who have "learned the task" (experts) [1, 8, 13, 24, 40, 42, 59].

The purpose of this review is to develop a hypothesis to explain why stick falls occur even for very highly skilled stick balancers. The discussion focuses on the development of stick balancing skill for seven subjects who had 10–85 days of daily stick balancing practice. The balance time provides a measurement of skill, i.e. the more skilled the stick balancer the longer on average that a stick of a given length can be balance on the fingertip. Our observations indicate that stick balancing skill increases with practice because the nervous system develops a feedback controller that takes advantage of an internal model. Thus, in some respects, the development of pole balancing skill is very similar to how the nervous system uses an internal model to acquire other voluntary motor skills [2, 23, 64]. However, as we demonstrate, control is challenged by the presence of sensory dead zones related to estimating the vertical displacement angle of the stick in the anterior-posterior direction and eye blinks. It is suggested that stick falls arise as a consequence of transient microchaos, a dynamical behavior that arises from the interplay between time delayed feedback, a sensory dead zone and the frequency-dependent encoding of force [22, 47].

2   Pole Balancing at the Fingertip

The stick was an oak dowel. The length of the stick (0.25, 0.39, 0.56, 0.91 m) was adjusted by cutting the dowel. Subjects were seated in a chair, faced a blank black screen and were required to keep their back against the back of the chair at all times while stick balancing on the fingertip of their dominant hand. Reflective markers were attached to both ends of the stick so that stick motions could be recorded by a motion capture system.

Skill level is measured as the average balance time, $\langle BT \rangle$, where BT denotes the balance time for a single balance trial. The *survival curve* shown in (Fig. 1a) is a plot of the fraction of trials still standing, i.e. the *fraction surviving*, as a function of consecutive days of practice [5, 6]. The survival curves can be described using a Weibull survival function [5, 32]

$$P(BT > t) \sim \exp[-(\lambda t)^{\beta}] \tag{1}$$

where $\lambda > 0$ and $\beta > 1$ are constants.

Fig. 1 Top: stick balancing
survival curves on different
days of practice. Bottom: the
time axis in the top panel is
re-scaled by replacing the
time, t, by $t/\langle BT \rangle$

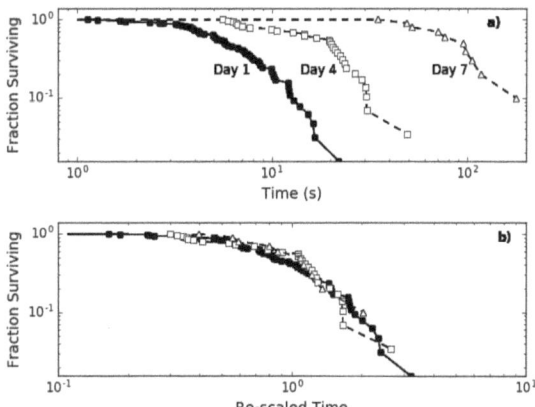

The more skilled the stick balancer the longer, on average, a stick can be balanced
at the fingertip. Thus with daily practice the survival curve shifts to the right. When
time, t, is re-scaled by $t/\langle BT \rangle$ the survival curves for different days collapse onto
a single survival curve (Fig. 1b). This observation indicates that $\langle BT \rangle$ is a relevant
time scale for the development of stick balancing skill [6]. Indeed, the increase
in $\langle BT \rangle$, estimated from practice sessions performed on consecutive days (≈ 20
trials each day) was typically greater than the increase in $\langle BT \rangle$ estimated from two
practice sessions performed on the same day (e.g. ≈ 20 trials on two separate
occasions; one in the morning, the other in the afternoon). This observation is
consistent with the known effects of sleep on motor skill acquisition [54, 61].

However, for the group of subjects studied in this communication, the balance
times quickly became so long that it was impractical to estimate $\langle BT \rangle$ in this
manner. For example, within 10 consecutive days of practice, all of these subjects
could balance a 0.56 m stick longer than 240 s! After 40 days of practice two subjects
could balance a 0.56 m stick on their fingertip for over 900 s! Thus if a subject was
able to balance a given stick length for 240 s for at least 1 out of 5 consecutive trials,
the subject is considered to have learned to balance a stick of this length.

3   Pendulum-Cart Model for Stick Balancing at the Fingertip

The starting point for modeling stick balancing at the fingertip is the pendulum-cart
model shown in Fig. 2. The corresponding governing equation reads [19, 20, 46]

$$
\begin{pmatrix} \frac{1}{3}m\ell^2 & \frac{1}{2}m\ell\cos\theta \\ \frac{1}{2}m\ell\cos\theta & m+m_0 \end{pmatrix} \begin{pmatrix} \ddot{\theta} \\ \ddot{x} \end{pmatrix} + \begin{pmatrix} -\frac{1}{2}mg\ell\sin\theta \\ -\frac{1}{2}m\ell\dot{\theta}^2\sin\theta \end{pmatrix} = \begin{pmatrix} 0 \\ F(t) \end{pmatrix}, \tag{2}
$$

where θ is the vertical displacement angle of the stick, ℓ is the stick length, m, m_0
are, respectively, the mass of the pole and cart, $\ddot{x}$ is the acceleration of the cart

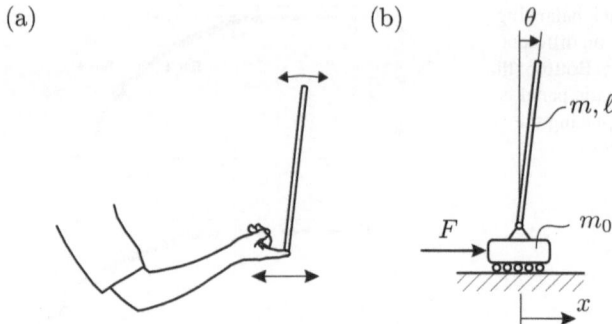

Fig. 2 Left: stick balancing on the fingertip. Right: the corresponding pendulum-cart model

(fingertip) and $F(t)$ describes the control force. The mass m_0 of the cart corresponds to the cumulate inertia of the moving arm segments of the subject (Fig. 2a).

Typically, human stick balancers spontaneously adopt a posture in which the wrist and fingers are held rigid so that the movements occur mainly at the elbow and shoulder. Thus, for balancing an oak stick on the fingertip, $m_0 \gg m$ (Fig. 2b). If the control force is zero, then elimination of the cyclic coordinate x and linearization around the upper fixed point yields

$$\ddot{\theta}(t) - \omega_n^2 \theta(t) = 0, \tag{3}$$

where $\omega_n = \sqrt{3g/2\ell}$ is the angular natural frequency of the pendulum hung downward.

The linearized equations of motion for the control of a pendulum-cart model are

$$\begin{pmatrix} \frac{1}{3}m\ell^2 & \frac{1}{2}m\ell \\ \frac{1}{2}m\ell & m+m_0 \end{pmatrix} \begin{pmatrix} \ddot{\theta} \\ \ddot{x} \end{pmatrix} + \begin{pmatrix} -\frac{1}{2}mg\ell & 0 \\ 0 & 0 \end{pmatrix} \begin{pmatrix} \theta \\ x \end{pmatrix} = \begin{pmatrix} 0 \\ F(t) \end{pmatrix}, \tag{4}$$

where x is the displacement of the fingertip from the starting point for pole (approximately half arm length when the subject is seated).

4 The Control Force

There are two considerations for choosing $F(t)$: (1) the state dependent variables are $\theta, \dot{\theta}$ and $\ddot{\theta}$ and (2) since the uncontrolled inverted pendulum position is unstable, it is not possible to use controllers such as the Smith predictor [37, 49] which assume that the only source of instability is the time delay [35, 36]. With this in mind there are two reasonable choices for the control force:

1. **Delayed state feedback:** An obvious choice is to use the most recently available values of $\theta(t-\tau), \dot{\theta}(t-\tau), \ddot{\theta}(t-\tau)$ and $x(t-\tau), \dot{x}(t-\tau), \ddot{x}(t-\tau)$, where τ is

the reaction time delay (closed-loop feedback delay). The proportional-derivative (PD) controller is

$$F_{PD}(t) = k_{p,\theta}\theta(t - \tau) + k_{d,\theta}\dot{\theta}(t - \tau) + k_{p,x}x(t - \tau) + k_{d,x}\dot{x}(t - \tau). \qquad (5)$$

where $k_{p,\theta}$, $k_{d,\theta}$, $k_{p,x}$, and $k_{d,x}$ are the proportional and derivative control gains for θ and x, respectively. With this choice of F, (4) becomes an example of a *retarded functional differential equation* (RFDE). The characteristic equation for RFDE's have infinitely many roots; however, instability arises when only a finite number of eigenvalues have positive real part [56]. If we include the possibility that $F(t)$ also depends on $\ddot{\theta}$ and $\ddot{x}$, then we have the proportional-derivative-acceleration (PDA) feedback and (4) becomes a neutral functional differential equation (NFDE).

2. **Predictor feedback:** Predictor feedback (PF) controllers suggest that rather than feeding back the delayed state, the controller should predict the actual state based on the known time history. Thus $F(t)$ is involved in making a prediction of the actual state variables [26]. Predictor feedback corresponds to a forward model in the neuroscience literature [34] and is often associated with finite spectrum assignment in the engineering control literature [26]. The system is asymptotically stable if all of the eigenvalues have negative real parts. Here we consider that the PF control, F_{PF}, has the form

$$F_{PF}(t) = \tilde{k}_{p,\theta}\theta(t - \tau) + \tilde{k}_{p,x}x(t - \tau) + \tilde{k}_{d,\theta}\dot{\theta}(t - \tau) + \tilde{k}_{d,x}\dot{x}(t - \tau)$$

$$+ \int_{t-\tau}^{t} k_f(t - s)f_{PF}(s)ds. \qquad (6)$$

The first four terms represent the delayed state feedback with control gains $\tilde{k}_{p,\theta}$, $\tilde{k}_{d,x}$, $\tilde{k}_{d,\theta}$, $\tilde{k}_{d,x}$ while the last term is associated with the weighted integral of the issued control force over the interval $[t - \tau, t]$. It has been shown that optimum prediction for a system with input delay is obtained by solving the system equations over the delay period [25, 33].

5   Feedback Identification

The important point is that if the time delay is known, then the feedback can be identified by measuring the shortest pole length, ℓ_{crit}, that can be balanced [20]:

- PD feedback: $\ell_{crit,PD} = 0.75g\tau^2$
- PDA feedback: $\ell_{crit,PDA} = 0.5\ell_{crit,PD}$
- PF control: $\ell_{crit,PF} = 0$

The time delay for human pole balancing at the fingertip is ≈ 220 ms. This value has been confirmed using three approaches: (1) the response following a visual blank produced using optical shutters [46], (2) from the response to a mechanical perturbation [34] and (3) from the stabilometry properties of pole balancing [48]. A time delay estimate of ≈ 100 ms was obtained using cross-correlation techniques [3]. However, the use of cross-correlation methods for estimating the time delay can be problematic [39, 51]. Here we use $\tau = 0.22$ s.

Figure 3 shows a plot of balance time versus pole length for expert pole balancers. The shortest pole that could be balanced for at least 1 out of 5 balance trials was ≈ 0.3 m. The vertical lines in the top panel give ℓ_{crit} using PD, PDA or PF control in the absence of a sensory dead zone. Since the human visual system is not very sensitive for detecting changes in acceleration [11], the observations in Fig. 3 suggest that expert pole balancers likely involves the development of an internal model.

The problem is that this analysis misses the point that even for the most skilled pole balancer the pole eventually falls. Indeed even when the subject succeeds in balancing a pole for 240 s on their fingertip, the very next balancing trial for the same pole length may last only a few seconds. This observation suggests that something important has been overlooked in the mathematical model.

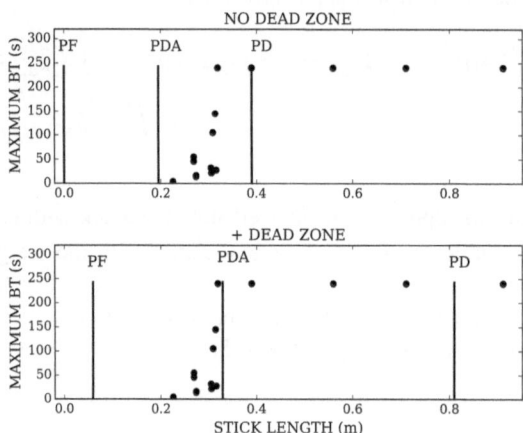

Fig. 3 Stick balance time (filled circle), BT, as a function of stick length, ℓ. Balancing was stopped if the stick had been balanced for 240 s for at least 1 out of 5 trials. For $\ell < 0.3$ m, the indicated balance time is the longest balance time of five trials. (top panel) The solid vertical lines indicate ℓ_{crit} calculated when there is no sensory dead zone. (bottom panel) The solid vertical lines show ℓ_{crit} estimated in the presence of a sensory dead zone [20]

6 Intermittent Forces

Considerable interest has focused on understanding the changes in velocity, ΔV, made by the fingertip as balancing skill increases [4, 8]. Measurements of ΔV provide insights into the nature of the forces generated by the movements of the fingertip to keep the stick balanced. Figure 4 shows the distribution of ΔV, $P(\Delta V)$, as a function of ℓ for a highly skilled subject with 85 days of practice (this is subject E1 in [46]). Clearly the shape of $P(\Delta V)$ changes as ℓ changes. For $\ell \geq 0.56$ m, $P(\Delta V)$ can be well approximated by a Gaussian distribution. In contrast, when $\ell < 0.56$ m $P(\Delta V)$ develops "tails" reminiscent of a Lévy distribution [4, 8]. The changes in the shape of $P(\Delta V)$ appear to have functional significance. For $\ell \leq 0.56$ m, this subject could balance 5/5 trials for 240 s. However, when $\ell = 0.39$ m the subject could balance only 2/5 trials longer than 240 s and when $\ell = 0.25$ m no trial lasted longer than 240 s. These observations suggest that a "tailed $P(\Delta V)$" is associated with less robust balance control.

Two explanations have been proposed to account for the tailed nature of $P(\Delta V)$. The first suggests that the changes in $P(\Delta V)$ arise from limitations in the control force [3]. In particular, it is suggested that the noisy fluctuations arise because an important control parameter has been noisily forced back and forth across a stability boundary. This mechanism is referred to as *on-off intermittency*. The presence of a $-3/2$ power law supports this interpretation. Moreover the control parameters measured for highly skilled stick balancers are tuned very close to the stability boundary [46]. The second suggests that the power laws are generated by an

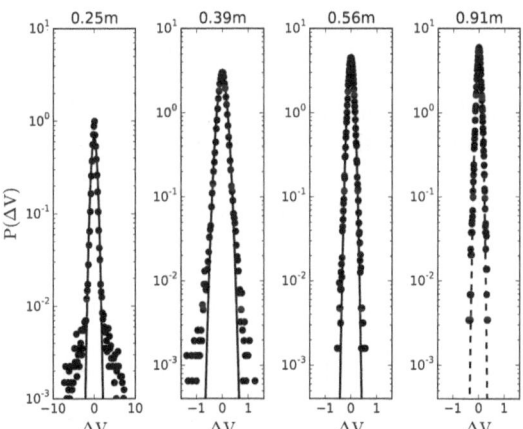

Fig. 4 Probability distributions for the changes in speed of the bottom marker during stick balancing at the fingertip as a function of stick length. The solid line shows a Gaussian distribution with the same standard deviation. The distributions for the 0.39, 0.5, and 0.91 m sticks were estimated from a single 240 s long trial of balancing data which was digitized at 1000 Hz. For the 0.25 m stick, the data was pooled from three consecutive balancing trials

intermittent control model that takes advantage of the flows in the neighbourhood of a saddle point [65].

7 Sensory Limitations

However, it is more likely that the difficulties encountered by a human stick balancers are related to sensory limitations. The initial motivation for this concept was the recognition that there is an inherent problem with continuous controllers in the presence of noise, namely, how to distinguish those fluctuations which need to be controlled from those that do not [43, 66]. This is because, by definition, there is a finite probability that an initial deviation away from a set point will be counter–balanced by one towards the set point just by chance. Too quick a response by a controller to a given deviation can lead to the phenomenon of "over control" leading to destabilization, particularly when time delays are appreciable. On the other hand, waiting too long runs the risk that the control may be applied too late to be effective. One way to overcome these problems is to use a switch–like, or discontinuous, feedback controller which is activated only when dynamical variables cross pre–set thresholds [43–45, 48].

There are two types on sensory limitations that can affect pole balancing. First, it is difficult for human subjects to estimate the vertical displacement angle of a pole balanced on the fingertip in the anterior-posterior (AP) direction. Several observations suggest that this sensory dead zone has functional significance: (1) > 70% of pole falls occur in the AP direction; (2) the amplitude of the fluctuations of θ in the AP direction are larger than those in the medial-lateral direction; (3) the presence of the dead zone is sufficient to account for the beneficial effects of vibration on pole balancing [20, 45] and (4) when a 0.9 m stick is balanced on a linear track oriented in the medial-lateral direction to eliminate the visual dead zone, then a PD controller is slightly better than a PF controller for maintaining stick balance [48]. Second, a natural occurring switching feedback occurs as a consequence of an eye blink. An eye blink lasts ≈ 0.2 s [62].

The presence of the dead zone necessitates a switching feedback, i.e. the feedback is turned on or off depending on whether θ_{AP} is larger or smaller than Π. This means that the angular position perceived by the neural system is

$$\theta_{\text{perceived}}(t - \tau) = \begin{cases} 0 & \text{if } |\theta_a(t - \tau)| < \Pi \\ \theta_a(t - \tau) & \text{if } |\theta_a(t - \tau)| \geq \Pi \ . \end{cases} \tag{7}$$

where θ_a is the stick's actual angle and Π is the functional sensory threshold We assume that information related to $\dot{\theta}$ and $\ddot{\theta}$ remains available [60]. The solid vertical lines in the bottom panel of Fig. 3 show ℓ_{crit} determined for PD, PDA and PF control when a sensory dead zone is included. Although the presence of the dead zone increases the shortest length of stick that can be balanced by a given feedback controller, it remains true that the stick balancer is most likely developing an internal model.

The positive effects of noise and vibration on balancing have been well documented previously: (1) noise can postpone the onset of instability in differential equation with delayed feedback control [31]; (2) standing on a vibrating platform improves stick balancing at the fingertip [45] and (3) wearing shoes with vibrating insoles improves balance in healthy elderly people [30].

8 Transient "Microchaos"

The presence of a sensory dead zone has profound effects on the dynamics of time-delayed dynamical systems [10, 12, 16, 47, 58]. In particular, the interaction between a sensory dead zone, time-delayed feedback and frequency-dependent discrete forces can generate "micro-chaos", i.e. a form of chaotic dynamics characterized by small amplitudes. These conditions are sufficient to satisfy the sufficient conditions for chaos [63]: (1) sensitivity to initial conditions, (2) the existence of closed invariant sets, and (3) mixing. When the variable is less that the sensory threshold there is no feedback. Thus the dynamics are unstable. Consequently the Lyapunov exponent is positive. The presence of negative feedback implies that invariant sets exist for appropriate choices of the parameters. Finally, the interplay between the limit cycle dynamics and the effects of time discretization on threshold crossings produce mixing. The role played by these simple considerations for producing microchaos have been verified analytically in simple models of time-delayed feedback [16, 22, 47].

However, the important point for our discussion is that adjacent to the region(s) in parameter space where long-lived micro-chaotic solutions exist, there is typically a region where transient micro-chaotic solutions arise [22, 27, 58]. Figure 5 compares the stick balancing survival curve for a novice stick balanced with $\langle BT \rangle = 31s$ to that generated by a micro-chaotic solution of (4) with switching feedback (7) (more details are given in the figure legend). There is surprisingly good agreement between the observed survival curve and those predicted by the model!

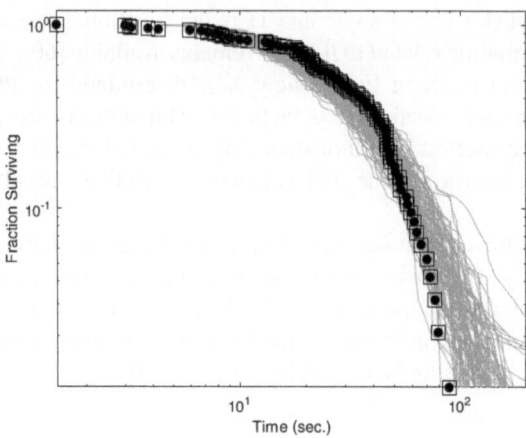

Fig. 5 Comparison of survival curves measured for a novice stick balancer (black symbols) to those predicted by (4) when $f(t)$ is given by (6) (gray lines). The survival curve for a novice stick balancer was determined from 95 balancing trials (black line) (mean balance time was 31 s). The balance times generated by this model are exquisitely sensitive to the choice of the initial stick balancing angle. This is because the model generates transient microchaos. A large pool of synthetic balance times was generated from the mathematical model by varying the initial angle in steps of 0.0003° to generate 33,333 "stick fall times". Each gray line in the figure represents the survival curve generated from 95 balance times randomly selected from this pool of 33,333 values. The gains were estimated using stabilometry [48] and numerical integration was performed using the semi-discretization method [21]. Parameters: $\tau = 0.24$ s, $\ell_s = 0.56$ m, $\ell_{arm} = 0.38$ m, $k_{p,\theta} = 120$ N/rad, $k_{d,\theta} = 28$ Ns/rad, $k_{p,x} = 14$ N/rad, $k_{d,x} = 19$ Ns/rad, $\Pi = 0.03$ rad. Note that ℓ_{arm} is less than the anatomical length of the arm measured from the top of the shoulder to the fingertip. This is because the fingertip movements become restricted by the bulk of the upper arm when the elbow is fully flexed

9   Concluding Remarks

Why does a stick balanced at the fingertip fall? This is obviously a difficult question to answer. It is easy to think of many possible reasons to account for a stick fall, for example, the human balancer may become distracted, the bottom of the stick slips off the fingertip, an eye blink occurs at a critical time, and so on. The surprising outcome of this review is that, taken together, the experimental observations fit quite nicely into the hypothesis that stick balancing dynamics exhibit transient microchaos. In other words, the falls are an intrinsic property of the control mechanism!

Of course we suspect that critics will say that all that this review has accomplished is to connect a complex process (stick falls) with a complex explanation (transient chaos). In our defence we point out that if we cannot understand how a stick balanced on a fingertip can fall, then how can we hope to understand why people fall? Transient micro-chaotic dynamics for stick balancing depend on three properties that can be readily assessed experimentally, namely, a sensory dead zone, time-delayed feedback and frequency-dependent encoding of force. It is certainly true that all of these factors are present for human balancing. Thus the more important question is whether an investigation of these three properties can provide insights into human falls. We leave this problem to future and younger researchers.

Acknowledgments Michael Mackey has been a dear friend and colleague of JM for over 47 years. We acknowledge that we have been able to learn from the experiences of many "pole balancing" collaborators including Juan Luis Cabrera (Universidad Politécnica de Madrid, Spain), Andre Longtin (University of Ottawa), Toru Ohira (Nagoya University, Japan), and from the Budapest University of Technology and Economics, Hungary (Gabor Csernak, Gabor Stepan and their students). In addition, JM acknowledges the 38 students at the Claremont Colleges who wrote senior theses on topics related to stick balancing and the 66 students who learned to stick balance. TI acknowledges support from the National Research, Development and Innovation Office (Grant No. NKFI-K138621).

References

1. Bilalić, M.: The Neuroscience of Expertise. Cambridge Fundamentals of Neuroscience in Psychology. Cambridge University Press, Cambridge (2017)
2. Bin, M., Huang, J., Isidori, A., Marconi, L., Mischiati abd E. Sontag, M.: Internal models in control, bioengineering and neuroscience. Annu. Rev. Control Robot. Auton. Syst. **9**, 718–727 (2022)
3. Cabrera, J.L., Milton, J.G.: On-off intermittency in a human balancing task. Phys. Rev. Lett. **89**, 158702 (2002)
4. Cabrera, J.L., Milton, J.G.: Human stick balancing: tuning Lévy flights to improve balance control. Chaos **14**, 691–698 (2004)
5. Cabrera, J.L., Milton, J.G.: Stick balancing: on-off intermittency and survival times. Nonlinear Stud. **11**, 305–31 (2004)
6. Cabrera, J.L., Milton, J.G.: Stick balancing, falls, and Dragon Kings. Eur. Phys. J. Spec. Top. **205**, 231–241 (2010)
7. Cabrera, J.L., Luciani, C., Milton, J.: Neural control on multiple time scales: insights from human stick balancing. Condens. Matter Phys. **9**, 373–383 (2006)
8. Cluff, T., Balasubramanian, R.B.: Motor learning characterized by changing Lévy distributions. PloS One **4**, e5998 (2009)
9. Courant, R., Robbins, F.: What is Mathematics? An Elementary Approach to Ideas and Method. Oxford University Press, Oxford (1941)
10. Csernak, G., Stepan, G.: Life expectancy of transient mictochaotic behavior. J. Nonlinear Sci. **15**, 63–91 (2005)
11. Dessing, J.C., Craig, C.M.: Bending it like Beckham: how to visually fool the goaltender. PLoS One **5**, e13161 (2010)
12. Enikov, E., Stepan, G.: Micro-chaos in digital control. J. Nonlinear Sci. **6**, 415–448 (1996)
13. Fitts, P.M., Posner, M.I.: Human Performance. Prentice/Hall International, London (1973)
14. Foo, P., Kelso, J.A.S., deGuzman, G.C.: Functional stabilization of fixed points: human pole balancing using time to balance. J. Exp. Psychol. Hum. Percept. Perform. **26**, 1281–1297 (2000)
15. Gawthrop, P., Lee, K.Y., Halaki, M., O'Dwyer, N.: Human stick balancing: an intermittency congtrol explanation. Biol. Cybern. **107**, 637–652 (2013)
16. Haller, G., Stepan, G.: Micro-chaos in digital control. J. Nonlinear Sci. **6**, 415–448 (1996)
17. Harrison, H.S., Kely-Stephen, D.G., Vaz, D.V., Michaels, C.F.: Multiplicative cascade dynamics in pole balancing. Phys. Rev. E **89**, 060903 (2014)

18. Insperger, T.: Act-and-wait concept for continuous-time control systems with feedback delay. IEEE Trans. Control Syst. Tech. **14**, 974–977 (2006)
19. Insperger, T., Milton, J.: Sensory uncertainty and stick balancing at the fingertip. Biol. Cybern. **108**, 85–101 (2014)
20. Insperger, T., Milton, J.: Delay and Uncertainly in Human Balancing Tasks. Lecture Notes on Mathematical Modeling in Human Balancing Tasks. Springer, Berlin (2021)
21. Insperger, T., Stepan, G.: Semi-discretization for Time-delay Systems: Stability and Engineering Applications. Applied Matheatical Sciences. Springer, Berlin (2011)
22. Insperger, T., Milton, J., Stepan, G.: Semi-discretization for time-delayed neural balance control. SIAM J. Appl. Dyn. Syst. **14**, 1258–1277 (2015)
23. Kawato, M.: Internal models for motor control and trajectory planning. Curr. Opin. Neurobiol. **9**, 718–727 (1999)
24. Kelso, J.A.S.: Dynamic Patterns: The Self-organization of Brain and Behavior. MIT Press, Cambridge (1999)
25. Kleinman, D.I.: Optimal control of linear system with time-delay and observational noise. IEEE Trans. Autom. Control **14**, 524–527 (1969)
26. Krstic, M.: Delay Compensation for Nonlinear, Adaptive, and PDE Systems. Birkhäuser, Boston (2009)
27. Lai, Y.C., Tél, Y.: Transient Chaos: Complex Dynamics on Finite-Time Scalles. Applied Mathematical Sciences, vol. 173. Springer, Berlin (2011)
28. Landry, M., Campbell, S.A., Morris, K., Aguilar, C.: Dynamics of an inverted pendulum with delayed feedback control. SIAM J. Appl. Dyn. Syst. **4**, 333–351 (2005)
29. Lee, K.Y., O'Dwyer, N., Halaki, M., Smith, R.: A new pardigm for human stick balancing: a suspended not an inverted pendulum. Exp. Brain Res. **221**, 309–328 (2012)
30. Lipsitz, L.A., Lough, M., Niemi, J., Travison, T., Howlett, H., Manor, B.: A shoe insole delivering subsensory vibrating insole improves balance and gait in healthy elderly people. Arch. Phys. Med. Rehabil. **96**, 432–459 (2015)
31. Longtin, A., Milton, J.G., Bos, J.E., Mackey, M.C.: Noise and critical behavior of the pupil light reflex at oscillation onset. Phys. Rev. A **41**, 6992–7005 (1990)
32. Mackey, M.C., Milton, J.G.: A deterministic approach to survival statistics. J. Math. Biol. **28**, 33–48 (1990)
33. Mannitus, A.Z., Olbrot, A.W.: Finite spectrum assignment for systems with delay. IEEE Trans. Autom. Control **AC-24**, 541–553 (1979)
34. Mehta, B., Schaal, S.: Forward models in visuomotor control. J. Neurophysiol. **88**, 942–953 (2002)
35. Miall, R.C., Jackson, J.K.: Adaptation to visual feedback delays in manual tracking: evidence against the Smith predictor model of human visually guided action. Exp. Brain Res. **34**, 543–551 (2006)
36. Miall, R., Weir, D.J., Wolpert, D.M.: Is the cerebellum a Smith predictor? J. Motor Beh. **25**, 203–216 (1993)
37. Michiels, W., Niculescu, S.I.: On the delay sensitivity of Smith predictors. Int. J. Syst. Sci. **34**, 543–551 (2003)
38. Milton, J.G.: Intermittent motor control: the "drift-and-act" hypothesis. In: Richardson, M.J., Riley, M., Shockley, K. (eds.) Progress in Motor Control: Neural, Computational and Dynamic Approaches, pp. 169–193. Springer, New York (2013)
39. Milton, J., Ohira, T.: Mathematics as a laboratory tool: dynamics, delays and noise, 2nd edn. Springer, New York (2021)
40. Milton, J.G., Small, S.L., Solodkin, A.: Neurophysiology of skilled performance. J. Clin. Neurophysiol. **21**, 133–227 (2004)
41. Milton, J.G., Small, S.S., Solodkin, A.: On the road to automatic: dynamic aspects in the development of expertise. J. Clin. Neurophysiol. **21**, 134–143 (2004)
42. Milton, J., Solodkin, A., Hluštík, P., Small, S.L.: The mind of expert motor performance is cool and focused. NeuroImage **35**, 804–813 (2007)

43. Milton, J.G., Cabrera, J.L., Ohira, T.: Unstable dynamical systems: delays, noise and control. Europhys. Lett. **83**, 48001 (2008)
44. Milton, J., Townsend, J.L., King, M.A., Ohira, T.: Balancing with positive feedback: the case for discontinuous control. Philos. Trans. R. Soc. A **367**, 1181–1193 (2009)
45. Milton, J.G., Ohira, T., Cabrera, J.L., Fraiser, R.M., Gyorffy, J.B., Ruiz, F.K., Strauss, M.A., Balch, E.C., Marin, P.J., Alexander, J.L.: Balancing and vibration: a prelude for "drift–and–act" balance control. PLoS One **4**, e7427 (2009)
46. Milton, J., Meyer, R., Zhvanetsky, M., Ridge, S., Insperger, T.: Control at stability's edge minimizes energetic costs: expert stick balancing. J. R. Soc. Interface **13**, 20160212 (2016)
47. Milton, J.G., Insperger, T., Cook, W., Harris, D.M. Stepan, G.: Microchaos in human postural balance: sensory dead zones and sampled time-delayed feedback. Phys. Rev. E **98**, 022223 (2018)
48. Nagy, D.J., Milton, J.G., Insperger, T.: Controlling stick balancing on a linear track: delayed state feedback or delay-compensating predictor feedback? Biol. Cybern. **117**, 113–127 (2023)
49. Palmor, Z.I.: Time delay compensation - Smith predictor and its modifications. In: Levine, W.S. (ed.) The Control Handbook, pp. 224–237. CRC and IEEE Press, Boca Raton (2000)
50. Puttemans, V., Wenderoth, N., Swinnen, S.P.: Changes in brain connectivity during the acquisition of a multifrequency bimanual coordinationtask: from the cognitive stage to advanced level of automaticity. J. Neurosci. **25**, 4270–4278 (2005)
51. Rubo, Z., Guanqun, L., Xueyao, L.: A time-delay estimation method against correlated noise. Proc. Eng. **23**, 445–450 (2011)
52. Schürer, F.: Zur theorie des balancierens. Math. Nache. **1**, 295–331 (1948)
53. Sieber, J., Krauskopf, B.: Extending the permissible control loop latency for the controlled inverted pendulum. Dyn. Syst. **20**, 189–199 (2005)
54. Siengsukon, C.F., Boyd, L.A.: Does sleep promote motor learning? Implications for physical rehabilitation. Phys. Therapy **89**, 370–383 (2009)
55. Srzednicki, R.: On periodic solutions in the Whitney's inverted pendulum problem. Discrete Contin. Dyn. Syst. Ser. S **12**, 2127–2141 (2019)
56. Stepan, G.: Retarded Dynamical Systems. Longman, Harlow (1989)
57. Stepan, G.: Delay effects in the human sensory system during balancing. Philos. Trans. R. Soc. A **367**, 1195–1212 (2009)
58. Stepan, G., Milton, J., Insperger, T.: Quantization improves stabilization of dynamical systems with delayed feedback. Chaos **27**, 114306 (2017)
59. Tenenbaum, G., Eklund, R.C.: Handbook of Sport Psychology. Wiley, Hoboken (2020)
60. Thiel, A., Greschner, M., Eurich, C.W.: Contribution of individual retinal ganglion cell responses to velocity and acceleration encoding. J. Neurophysiol. **98**, 2285–2296 (2007)
61. Walker, M.P., Brakefield, T., Seidman, J., Hobson, J.A., Stickgold, R.: Sleep and the time course of motor skill learning. Learn. Mem. **10**, 275–284 (2003)
62. Wang, Y., Toon, S.S., Gautam, R., Henson, D.B.: Blink frequency and duration during the perimetry and their relationship to test-retest threshold variability. Inv. Ophthal. Visual Sci. **52**, 4546–4550 (2011)
63. Wiggins, S.: Chaotic Transport in Dynamical Systems. Springer, New York (1992)
64. Wolpert, D.M., Ghahramani, Z.: Computational principles of movement neuroscience. Nat. Neurosci. Suppl. **3**, 1212–1217 (2000)
65. Yoshikawa, N., Suzuki, Y., Kiyono, K., Nomura, T.: Intermittent feedback-control strategy for stabilizing inverted pendulum on manually controlled cart as analogy to human stick balancing. Front. Comput. Neurosci. **10**, 34 (2016)
66. Zgonnikov, A., Lubashevsky, I., Kanemoto, S., Miyazawa, T., Suzuki, T.: To react or not to react? Intrinsic stochasticity of human control in virtual stick balancing. J. R. Soc. Interface **11**, 20140236 (2014)